Directed Mutagenes

Maria B. Red
30.12.91, Blackwells,
Oxford

The Practical Approach Series

SERIES EDITORS

D. RICKWOOD

Department of Biology, University of Essex
Wivenhoe Park, Colchester, Essex CO4 3SQ, UK

B. D. HAMES

Department of Biochemistry and Molecular Biology, University of Leeds
Leeds LS2 9JT, UK

Affinity Chromatography
Animal Cell Culture
Animal Virus Pathogenesis
Antibodies I and II
Biochemical Toxicology
Biological Membranes
Biosensors
Carbohydrate Analysis
Cell Growth and Division
Centrifugation (2nd edition)
Clinical Immunology
Computers in Microbiology
Directed Mutagenesis
DNA Cloning I, II, and III
Drosophila
Electron Microscopy in Molecular Biology
Fermentation
Flow Cytometry
Gel Electrophoresis of Nucleic Acids (2nd edition)
Gel Electrophoresis of Proteins (2nd edition)
Genome Analysis
HPLC of Small Molecules

HPLC of Macromolecules
Human Cytogenetics
Human Genetic Diseases
Immobilised Cells and Enzymes
Iodinated Density Gradient Media
Light Microscopy in Biology
Liposomes
Lymphocytes
Lymphokines and Interferons
Mammalian Development
Medical Bacteriology
Medical Mycology
Microcomputers in Biology
Microcomputers in Physiology
Mitochondria
Mutagenicity Testing
Neurochemistry
Nucleic Acid and Protein Sequence Analysis
Nucleic Acids Hybridisation
Nucleic Acid Sequencing
Oligonucleotide Synthesis

Directed Mutagenesis

A Practical Approach

Edited by

M. J. McPHERSON

Department of Biochemistry and Molecular Biology,
University of Leeds,
Leeds LS2 9JT, UK

OXFORD UNIVERSITY PRESS
Oxford New York Tokyo

Oxford University Press, Walton Street, Oxford OX2 6DP

Oxford   New York   Toronto
Delhi   Bombay   Calcutta   Madras   Karachi
Petaling Jaya   Singapore   Hong Kong   Tokyo
Nairobi   Dar es Salaam   Cape Town
Melbourne   Auckland

and associated companies in
Berlin   Ibadan

Oxford is a trade mark of Oxford University Press

Published in the United States
by Oxford University Press, New York

British Library Cataloguing in Publication Data
Directed mutagenesis.
1. Mutagenesis
I. McPherson, M. J.
575.292
ISBN 0–19–963141–7
ISBN 0–19–963140–9 (pbk)

Library of Congress Cataloging in Publication Data
Directed mutagenesis : a practical approach / edited by M. J.
McPherson.
p.   cm. — (The Practical approach series)
Includes bibliographical references and index.
1. Site-specific mutagenesis.   I. McPherson, M. J.   II. Series.
QH465.5.D57 1991   574.87'3282—dc20   90–46788
ISBN 0–19–963141–7
ISBN 0–19–963140–9 (pbk)

Set by
Footnote Graphics, Warminster, Wilts

Printed in Great Britain by
Information Press Ltd, Oxford

Preface

Mutagenesis provides an essential tool for the dissection of structure and function relationships in biological systems. The title of this volume, '*Directed mutagenesis*', reflects our ability to target precisely the location and type of modification we wish to introduce into a cloned DNA sequence. At its most accurate, directed mutagenesis allows the change of a single specified base to any of the other three bases. It also allows the introduction of deletions and insertions with predefined end-points. Such a precise ability to engineer changes results from the use of oligonucleotide mutagens which are now widely available to molecular biology laboratories at relatively low cost. However, directed mutagenesis is not restricted to primer-directed procedures but covers many other approaches for the targeted alteration of the base sequence of cloned DNA.

It may seem paradoxical to bring together methods that, on the surface, one could consider to be competitors for the attention of the potential mutagenesis exponent. However, the approaches presented in this volume are often complementary and may be used in combination for the detailed dissection and manipulation of DNA. The chapters are written by experts who, in most cases, have been closely involved in the conception and methodological development of the approach about which they write.

The aim has been to provide a text that will suit both the beginner and the expert. Each chapter provides information on experimental design, protocols for performing mutagenesis experiments and, in many cases, information to help assess how well an experiment has worked, or why it has not worked! For those who already use mutagenesis procedures the volume should provide updates of favoured approaches, new and tested methods, a comparison of alternative protocols and will hopefully stimulate new ideas for mutagenesis experiments.

Oligonucleotide primer-directed mutagenesis experiments were pioneered by Mark Zoller and Michael Smith who laid the foundations for the development of the widely used M13-based mutagenesis procedures. The current use of such an approach for precise and rapid mutagenesis stems from the development of methods for enhancing the recovery of the desired mutational variant. Three such high-efficiency mutagenesis methods are provided in Chapters 1 to 3. These approaches, which rely on an ability to select against the wild-type or for the mutant DNA strand, involve restriction-selection or restriction-purification coupled with the use of repair-deficient host strains (Chapter 1), the use of uracil-containing template DNA (Chapter 2), or the incorporation of phosphorothioate nucleotides (Chapter 3). All are easy to perform and allow identification of the correct mutant by direct DNA sequencing of a small number of clones (usually 2 to 4). As cloning vectors have developed so they have been co-opted for use in mutagenesis experiments. Thus, many of the procedures outlined in this volume make use of 'phagemid' vectors which combine the advantages of plasmids and of single-stranded phage. Such vectors are small, encode resistance to an antibiotic, and can be manipulated as multicopy plasmids. In addition phagemids carry a replication origin from a single-stranded phage allowing rescue of single-stranded DNA for both mutagenesis and DNA sequencing.

In cases where M13 or phagemid vectors are not suitable for expression of a gene it can be advantageous to perform mutagenesis of a double-stranded plasmid. This prevents tedious and time-consuming transfer of the gene between a single-stranded 'mutagenesis' vector and the parent 'expression' vector. Two such approaches are presented, the first involves the production of a gapped-heteroduplex substrate for mutagenesis (Chapter 4) while the second, described in Chapter 5, uses the phosphorothioate chemistry introduced in Chapter 3.

Information from biochemical, chemical, sequence comparison, and structural studies can help define specific nucleotides as targets for mutagenesis. However, where such information is limited, a random mutagenesis strategy can be applied where there are no preconceived notions about which nucleotides or codons represent important targets. Random in this context means the directed mutagenesis of a defined region of DNA in such a way as to isolate a series of mutations dispersed throughout the target sequence. Isolation of interesting mutational variants often involves some form of functional selection while characterization of the variants can lead to the identification of sequences which warrant detailed investigation. A number of random mutagenesis approaches are presented in Chapters 6 to 10.

Linker-scanning mutagenesis can introduce a cluster of point mutations at random positions throughout a sequence by the replacement of a few consecutive bases with linkers of equivalent size and which can be topologically identified (Chapter 6). The power of chemical mutagens has been know for many years and their use for inducing point mutations is coupled, in Chapter 7, with a denaturing gel system for distinguishing between mutant and wild type sequences.

Forced misincorporation of nucleotides (Chapter 8) and the use of 'spiked' oligonucleotides as mutagenic primers (Chapter 9) are particularly useful for generating very large populations ($>10^5$) of mutations of complete genes. These two procedures also make use of methods, described in Chapters 2 and 3 respectively, for destroying the wild-type template DNA to enhance the recovery of mutants.

Cassette mutagenesis (Chapter 10) can be used to introduce random mutations but is also widely used for the detailed study of part of a gene. In this approach synthetic oligonucleotides, which include the desired mutation(s), are designed to form a cassette to replace a corresponding segment of the wild-type gene. Finally, the polymerase chain reaction has emerged as a powerful tool not only for the analysis and cloning of DNA sequences but also, as described in Chapter 11, for facilitating their recombination and mutagenesis.

As with any volume endeavouring to cover such an exciting and evolving field as directed mutagenesis this text represents a methodological snap-shot in which it is impossible to cover every mutagenic technique. I hope this volume provides a valuable and representative 'users' guide to current directed mutagenesis practice. The authors are all leading experts in the field of directed mutagenesis and I am grateful to them for their time and effort during this project. I would also like to thank the publishers for their help during the preparation of this volume and my colleagues, especially Sarah Gurr, for help and encouragement. My special thanks go to Sue and Andrew for their understanding and patience.

Leeds   M.J.M.
October 1990

Contents

7. Random chemical mutagenesis and the non-selective isolation of mutated DNA sequences in *vitro*   135

C. Walton, R. K. Booth, and P. G. Stockley

8. An enzymatic method for the complete mutagenesis of genes

J. K. C. Knowles and P. Lehtovaara

9. Spiked oligonucleotide mutagenesis

S. C. Blacklow and J. R. Knowles

Contents

Appendix

Index

Contributors

S. C. BLACKLOW
Department of Chemistry, Harvard University, Cambridge, MA 02138, USA.

R. K. BOOTH
Department of Genetics, University of Leeds, Leeds, LS2 9JT, UK.

P. CARTER
Department of Protein Engineering, Genentech Inc., 460 Point San Bruno Boulevard, South San Francisco, CA 94080, USA.

F. ECKSTEIN
Max-Plank-Institut für Experimentelle Medizin, Abteilung Chemie, Herman-Rein Strasse 3, D-3400 Gottingen, FRG.

J. GEISSELSODER
Bio-Rad Laboratories, Chemical Division, Molecular Biology Products Group, 3300 Regatta Way, Richmond, CA 94804, USA.

R. M. HORTON
Department of Immunology, Mayo Clinic, Rochester, MN 55905, USA.

M. INOUYE
Department of Biochemistry, Robert Wood Johnson Medical School, University of Medicine and Dentistry of New Jersey, Piscataway, NJ 08854, USA.

S. INOUYE
Department of Biochemistry, Robert Wood Johnson Medical School, University of Medicine and Dentistry of New Jersey, Piscataway, NJ 08854, USA.

J. K. C. KNOWLES
Glaxo Institute for Molecular Biology S.A., Route des Acacias 46, 1211 Geneva 24, Switzerland.

J. R. KNOWLES
Department of Chemistry, Harvard University, Cambridge, MA 02138, USA.

P. LEHTOVAARA
VTT Biotechnical Laboratory, SF 02150 Espoo, Finland.

B. LUCKOW
Institut für Zell- und Tumorbiologie, Deutsches Krebsforschungszentrum, Im Neuenheimer Feld 280, D-6900 Heidelberg 1, FRG.

J. McCLARY
Bio-Rad Laboratories, Chemical Division, Molecular Biology Products Group, 3300 Regatta Way, Richmond, CA 94804, USA.

D. B. OLSEN
Max-Plank-Institut für Experimentelle Medizin, Abteilung Chemie, Herman-Rein Strasse 3, D-3400 Göttingen, FRG.

L. R. PEASE
Department of Immunology, Mayo Clinic, Rochester, MN 55905, USA.

J. H. RICHARDS
Division of Chemistry and Chemical Engineering, California Institute of Technology, Pasadena, CA 91125, USA.

J. R. SAYERS
Max-Plank-Institut für Experimentelle Medizin, Abteilung Chemie, Herman-Rein Strasse 3, D-3400 Göttingen, FRG.

G. SCHUTZ
Institut für Zell- und Tumorbiologie, Deutsches Krebsforschungszentrum, Im Neuenheimer Feld 280, D-6900 Heidelberg 1, FRG.

P. G. STOCKLEY
Department of Genetics, University of Leeds, Leeds, LS2 9JT, UK.

C. WALTON
Department of Genetics, University of Leeds, Leeds, LS2 9JT, UK.

F. WITNEY
Bio-Rad Laboratories, Chemical Division, Molecular Biology Products Group, 3300 Regatta Way, Richmond, CA 94804, USA.

P. D. YUCKENBERG
Bio-Rad Laboratories, Chemical Division, Molecular Biology Products Group, 3300 Regatta Way, Richmond, CA 94804, USA.

Abbreviations

AMV	avian myeloblastosis virus
ATP	adenosine triphosphate
BAP	bacterial alkaline phosphatase
Ci	Curie
DGGE	denaturing gradient gel electrophoresis
DMSO	dimethyl sulphoxide
DNase	deoxyribonuclease
dNMPαS	deoxythionucleotide
dNTP	deoxynucleotide triphosphate
ds	double-stranded
DTT	dithiothreitol
EDTA	ethylenediaminetetraacetate
HPLC	high performance liquid chromatography
IPTG	isopropyl-β-D-thiogalactoside
Km^R	kanamycin resistance
LS	linker scanning
Met	methionine
m.o.i.	multiplicity of infections
PBS	phosphate-buffed saline
PCR	polymerase chain reaction
PEG	polyethylene glycol
p.f.u.	plaque-forming unit
RF	replicative form
RNase	ribonuclease
r.t.	room temperature
TBE	Tris-borate-EDTA
TE	Tris–EDTA
TEMED	N,N,N',N'-tetramethylenediamine
TFB	transformation buffer
TLC	thin-layer chromatography
SDS	sodium dodecyl-sulphate
ss	single-stranded
T_m	melting temperature
Tris	tris(hydroxy methyl)amino methane
UV	ultraviolet
X-gal	5-bromo-4-chloro-3-indolyl-β-D-galactopyranoside

Mutagenesis facilitated by the removal or introduction of unique restriction sites

PAUL CARTER

1. Introduction to site-directed mutagenesis

Over the last decade, a variety of efficient and reliable methods have been developed for the construction of site-directed mutations in DNA using synthetic oligonucleotides (reviewed in refs 1–3). This has revolutionized the study of gene regulation and protein structure and function. All of these mutagenesis methods fall into one of three broad strategies. One approach uses an oligonucleotide complimentary to part of a single-stranded (ss) DNA template but containing an internal mismatch(es) to direct the desired mutation (see Section 1.1, and Chapters 2 to 5 and 9). A second strategy is to replace the region of interest with a synthetic mutant fragment generated by annealing complimentary oligonucleotides ('cassette mutagenesis' (ref. 4 and Chapter 10)), or by hybridization and ligation of a number of oligonucleotides (reviewed in ref. 5). The third and most recent strategy, has been to harness the power of the polymerase chain reaction (PCR) (6) to generate a mutant fragment starting from a double-stranded (ds) DNA template using mismatched oligonucleotides (refs 7 and 8, and Chapter 11). In this chapter an efficient method of site-directed mutagenesis using mismatched oligonucleotides (first strategy above) is discussed.

1.1 Mutagenesis using mismatched oligonucleotides

A general scheme for the construction of mutants using mismatched oligonucleotides is shown in *Figure 1*. The mutagenic oligonucleotide is first phosphorylated ('kinased') at the 5′-end with T4 polynucleotide kinase and then annealed to a ss DNA template which may be either M13 or a ss plasmid ('phagemid'; see Section 1.4). The primer is then extended in the presence of a DNA polymerase and deoxynucleoside triphosphates (dNTPs) and the ends of the nascent strand joined wth T4 DNA ligase. The polymerase used is often the Klenow fragment of *Escherichia coli* DNA polymerase I ('Klenow'),

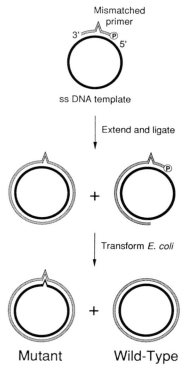

Figure 1. Mutagenesis using mismatched oligonucleotides (see Section 1.1). A mismatched and phosphorylated oligonucleotide primer is annealed to a ss DNA template (in this case a phagemid) and extended in the presence of dNTPs and Klenow and also T4 DNA ligase to join the ends of the nascent strand. The mixture of fully and partially ds heteroduplex DNA is transformed into an *E. coli* host and gives rise to mutant and wild-type daughter molecules.

which lacks the 5'→3' exonuclease activity of the holo enzyme. However, other enzymes such as T4 DNA polymerase (9, 10) and T7 DNA polymerase (11) are becoming increasingly widely used since they offer a number of advantages over the use of Klenow (see Section 1.2). The heteroduplex DNA is transformed into an *E. coli* host strain where it may give rise to mutant and wild-type progeny. This approach is very versatile and in addition to simple point mutations it may be used to construct any combination of point mutations, insertions and deletions.

1.2 Factors reducing yield

A number of factors in the *in vitro* construction of heteroduplex DNA may act to reduce mutant yield (reviewed in refs 1–3):

(a) The mutagenic oligonucleotide may prime inefficiently at the target site or spuriously at additional sites. These problems can often be overcome by the judicious design of the mutagenic oligonucleotide and by titrating the ratio of primer to template (see Sections 1.5 and 3.3).

(b) Klenow may displace the mutagenic primer after extending all-the-way-around the template. However, ligation may be promoted over the competing displacement reaction by using high ratios of ligase to Klenow, by avoiding A/T rich stretches at the 5'-end of the mutagenic primer and by performing the extension/ligation reactions at low temperature (10–14°C). This problem may also be alleviated by the use of alternative DNA polymerases which are less prone to displacement such as T4 or T7 DNA polymerase.

(c) Small RNA molecules (from lysed cells) present in the template may prime randomly. This 'self-priming' activity of the template can be virtually eliminated by digestion of contaminating RNA with ribonuclease A (12).

(d) Trace amounts of dUTP present (from oxidative deamination of dCTP) in the dNTPs used in the *in vitro* DNA synthesis reaction results in deoxyuridine being incorporated into the nascent strand in place of deoxythymidine. This can result in strand cleavage and nick translation *in vivo* which may reduce mutant yield (13). This problem can be alleviated by using high quality (HPLC purified) dNTPs.

(e) Inefficient DNA synthesis may occur as a consequence of Klenow stalling at regions of extensive secondary structure in the template. T4 gene 32 protein which binds to ss DNA has been used in conjugation with T4 DNA polymerase in an attempt to overcome this problem (9), and may give at least a modest improvement in mutant yield compared to the use of Klenow (10). T7 DNA polymerase has several advantages over Klenow for use in heteroduplex construction in that it is highly processive (synthesizes thousands of nucleotides without dissociating from the template) and yet is not prone to strand displace. Furthermore, heteroduplex DNA may be efficiently constructed in a few hours and transformed into *E. coli* the same day. This is in contrast to Klenow where overnight incubations generally give better results than shorter reaction times (12). Native T7 DNA polymerase is preferable to a chemically modified version, Sequenase® [developed for dideoxynucleotide sequencing (14)] for heteroduplex construction. This is because Sequenase® lacks the 3'→5' exonuclease (proof-reading) activity decreasing its fidelity and increasing the risk of unwanted mutations as a result of nucleotide misincorporation. Furthermore, native T7 DNA polymerase is less prone to cause strand displacement than the chemically modified version.

After transformation of the heteroduplex DNA into *E. coli* several additional factors may lower the recovery of mutants:

(a) One of the several mismatch repair systems identified in *E. coli* is directed by methylation of GATC sites by the *dam* methylase (see ref. 15 and additional literature cited therein). After transfection of *E. coli* with heteroduplex DNA, the *dam* system will direct mismatch repair towards the template strand, which is at least partially methylated, and thus remove the mutation (16, 17).

(b) Progeny may be derived from either strand of the heteroduplex DNA. In the case of M13 (but not phagemids) there is a 2:1 bias in favour of the minus (mutagenized) strand which reflects the asymmetric nature of M13 replication (16, 18).

(c) If the ends of the nascent strand in the heteroduplex are not ligated then the mismatches can potentially be removed by 5′→3′ exonuclease activity *in vivo*. The conversion of ss template into covalent closed circular heteroduplex DNA is often rather inefficient: in one case, on examining the products of 19 different extension/ligation reactions <30% of the newly synthesized DNA was found in a covalent closed circular form (19).

These various problems result in low and variable yields of mutants which in the past necessitated the use of a screen to identify mutants (e.g. by hybridization with the mutagenic oligonucleotide or by restriction digestion) prior to verification by nucleotide sequencing. However, a number of high efficiency site-directed mutagenesis methods are now available (reviewed in refs 1–3), including the one described in this chapter and those in Chapters 2 and 3. These methods have largely overcome the difficulties described above and so largely obviated the need to screen for mutants. Mutagenesis kits representing several high efficiency mutagenesis methods are commercially available including those of Kunkel (ref. 13, and Chapter 2) from Bio-Rad, Eckstein (ref. 20, and Chapter 3) from Amersham, and Batt (21) from United States Biochemicals.

1.3 Strategies to enhance mutant yield

Reduction in mutant yield by mismatch repair *in vivo* (see Section 1.2) has prompted the use of *E. coli* host strains which are deficient in point mismatch repair (e.g. containing *mutL* or *mutS* mutations (16)) to enhance mutant yield (12, 22, 23). The recovery of mutants may be increased by up to 10-fold by simply transforming the heteroduplex DNA into a repair deficient (*mutL*) *E. coli* host strain instead of a repair proficient (but otherwise isogenic) strain (12).

The use of repair deficient host strains has been combined with the use of genetic selection to eliminate progeny derived from the wild-type strand of heteroduplex DNA using the 'gapped-duplex' (22, 23) and 'coupled priming' techniques (12). In the coupled priming technique, one primer is used to direct the silent mutation of interest and a second primer is used to remove a selectable marker in the template such as a stop (e.g. amber) codon in an essential gene or an *Eco*K or *Eco*B recognition site. The resultant hetero-duplex DNA is then transformed into a repair deficient *E. coli* strain which also selects against wild-type progeny (e.g. a non-suppressor strain in the case where an amber marker is used), to increase the yield of mutants by up to 20-fold. Although this strategy works well (12), further improvement would allow mutants to be identified directly by nucleotide sequencing, without first having to screen by oligonucleotide hybridization.

Substantial enrichment for mutants may be achieved by the procedures of 'restriction-selection' and 'restriction-purification' (24). In designing a primer to mutate a gene coding for some protein of interest, it is almost invariably possible to install or remove a unique restriction site in close proximity to the desired mutation without changing the amino acid sequence (i.e. 'silent muta-tion'; see Section 1.5). If a unique restriction site is removed by mutagenesis then wild-type DNA may be selectively eliminated from the mixed pool of wild-type and mutant DNA by digestion with the corresponding endonuclease (restriction-selection; *Figure 2*). Alternatively, if a unique restriction site is introduced during the mutagenesis reaction, mutant DNA may be purified away from the wild-type DNA by digestion and isolation of the linearized DNA on a polyacrylamide gel (restriction-purification; *Figure 2*). The mutant DNA is recovered by electroelution, self-ligation, and transformation of *E. coli*.

In this chapter, the use of repair deficient host strains has been combined with restriction-selection and restriction-purification to enhance the utility and reliability of these procedures for oligonucleotide-directed mutagenesis. Protocols are provided for use with phagemid vectors which have largely superceded M13 vectors (see Section 1.4). However restriction-selection and restriction-purification should also be applicable to M13 (e.g. by combining with protocols for mutagenesis in M13 vectors described in ref. 25). It is advisable to remove any contaminating ss DNA in preparations of M13 ds DNA (e.g. by filtration through nitrocellulose filters as described in Chapter 3 (Section 2.4) and ref. 20) since this may otherwise interfere with restriction-selection and restriction-purification.

1.4 Phagemid vectors

In the past, mutations were commonly constructed and sequence verified using M13 vectors, where ss DNA could be very readily obtained. Sub-sequently mutant genes usually had to be subcloned into suitable plasmid expression vectors. Nowadays the subcloning step can be obviated by the use of phagemid vectors, (also known as 'phasmids' or 'plage') which combine the advantages of plasmids and phage (26). Phagemids are plasmids which contain an origin of DNA replication from a filamentous bacteriophage such as M13 or f1. Upon infection with a 'helper' phage of cells harbouring phagemid DNA, one strand of phagemid DNA is synthesized, packaged into

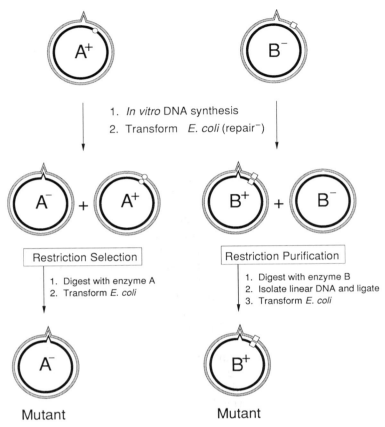

Figure 2. Mutagenesis facilitated by using a repair deficient host strain and restriction-selection or restriction-purification (see Section 1.3). Heteroduplex DNA is constructed (as in *Figure 1*) and transformed into a repair deficient (repair⁻) *E. coli* host strain. The mutagenic primer may be synthesized to remove a unique restriction endonuclease site (A⁺→A⁻, ○) allowing restriction-selection with restriction enzyme A to eliminate wild-type DNA. Alternatively the mutagenic primer may be designed to introduce a unique restriction endonuclease site (B⁻→B⁺, □) allowing restriction-purification of mutagenized DNA using restriction enzyme B.

pseudo phage particles and secreted into the media (27). An important advance in the development of phagemids has been the construction of suitable helper phage such as M13K07 (28) which promotes preferential packaging of ss phagemid DNA over ss phage DNA.

1.5 Design of mutagenic oligonucleotide primers

The length and location of mismatches are usually the most important considerations in the design of mutagenic oligonucleotides. There is a great deal of flexibility for personal choice of these parameters; however, the reliability and expense of locally available oligonucleotides should be taken into account. Use 8–13 perfectly matched nucleotides at each end of the mutagenic primer for simple or multiple point mutations. For more complicated replacements such as insertions, deletions, or any combination of insertion, deletion, and point replacements, use 14–20 perfectly matched residues at each end of the primer. The maximum working limit of oligonucleotide length is at least 100 nucleotides (see *Figure 3*) and apparently is limited by the reliability of synthesis and purification of very long oligonucleotides. As oligonucleotides are usually synthesized by stepwise addition of nucleotide monomers, the final yield will naturally decrease with oligonucleotide length and at some point will become limiting. Furthermore, the longer the oligonucleotide the more difficult it becomes to resolve full length product from unwanted side-products (e.g. where monomer coupling has failed at one cycle) by polyacrylamide gel electrophoresis.

Computer software is commercially available to 'reverse translate' a protein sequence into degenerate nucleotide sequences to assist in the design of oligonucleotides which install or remove restriction sites. Criteria for choosing between several different possible restriction endonucleases are discussed in Section 1.6 below. The entire nucleotide sequence of the construction to be mutagenized may not be known. In this case, test digests may be performed to verify the presence or absence of a particular restriction site.

Computer programs are also available to check the proposed oligonucleotide sequence for homology with the template. This is not generally necessary; however, it is advisable to avoid a stretch of residues at the 3'-end of the oligonucleotide which are perfectly matched to a non-target site on the

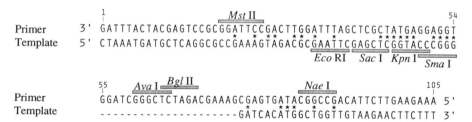

Figure 3. Example of a complex mutation which was assisted by restriction-selection. The 105-mer oligonucleotide shown was used to insert 21 nucleotides (– –) and make numerous point changes (★) in the template. The unique *EcoR*I site in the template was removed during mutagenesis and was used for restriction-selection. The template also contains unique sites for *Sac*I, *Kpn*I and *Sma*I (derived from a polylinker) which represent alternative sites for restriction-selection. The mutagenic primer introduces new and unique sites for *Mst*II and *Ava*I to assist in subsequent cloning experiments. However these sites could also have been used to enrich for mutants by restriction-purification. One completely correct clone was found among the first six clones sequenced—the others were all wild-type (see Section 3.1).

template. It is also prudent to avoid (if possible) A/T rich stretches at the 5'-end of the oligonucleotide to minimize possible displacement by Klenow in heteroduplex construction (see Section 1.2). Coupling between the silent target mutation and the restriction site change in repair deficient host strains is generally very tight and largely independent of whether the mismatch(es) for the restriction site change are on the 5'- or 3'-side of the silent target mutation. The extent of coupling becomes more relaxed as the separation between these two sites is increased (see Section 3.1 (b)) and is most effective for separations of less than 30 nucleotides.

1.6 Choice of endonuclease for restriction-selection and restriction-purification

Many different restriction endonucleases have been used successfully for restriction-selection and restriction-purification; however, some enzymes undoubtedly work considerably better than others. Some properties of restriction enzymes significantly affect their use in the mutant enrichment procedures (see manufacturers' catalogues e.g. New England Biolabs, for details):

(a) In many strains of *E. coli* the N^6 of adenine in the sequence GATC is methylated by the *dam* methylase (see Section 1.2) and the internal cytosine in the sequences CCAGG and CCTGG is methylated by the *dcm* methylase. Several restriction enzymes will not cleave at their normal recognition site if it is methylated e.g. cleavage by *ClaI* is blocked by *dam* methylation at an overlapping GATC site. If necessary the ds DNA pool after mutagenesis may be passaged through a *dam* minus or *dcm* minus strain as appropriate, to allow cleavage with a particular enzyme. If possible it is usually more convenient to choose a different restriction enzyme for enrichment so that this passaging step is not necessary.

(b) Some enzymes such as *Eco*RI, *Bam*HI and *Hin*dIII show altered substrate specificity ('star activity') when the digestion conditions are changed; for example, at low ionic strength or at high pH or in the presence of high concentrations of glycerol or DMSO. In utilizing these enzymes for restriction-selection or restriction-purification, use reaction conditions recommended by the manufacturer to avoid star activity.

If one has several potential restriction enzymes then a good choice would be an enzyme where available batches cut reliably and quantitatively and are free of contaminating exonuclease activity. For restriction-selection (but not restriction-purification) it is essential to ensure that the digest proceeds as close to completion as possible by using relatively small amounts of DNA and large amounts of the endonuclease (see Section 2.6.3). It is desirable to choose an enzyme which is cheap and available in a concentrated form (5–25 units/μl). For restriction-purification it is preferable (but not essential) to choose an enzyme which cleaves to leave staggered ('sticky') rather than blunt ends, because religation is more efficient.

1.7 Methods included in this chapter

My aim has been to include enough detail in this chapter to allow the reader to construct and verify mutants in phagemid vectors and to provide reliable protocols which may be combined with other mutagenesis methods or techniques in molecular biology. Several important procedures are not described because they represent basic techniques for which reliable protocols have been published elsewhere, including other chapters in this book: preparation and transformation of $CaCl_2$-competent cells (25, 29, 30), preparations of ds DNA ('minipreps', see refs 29 and 30) and dideoxy nucleotide sequencing (31). Detailed protocols for the synthesis of oligonucleotides (32) and their purification by polyacrylamide gel electrophoresis (30, 32, 33) are also available.

2. Oligonucleotide-directed mutagenesis facilitated by using repair deficient host strains and restriction-selection or restriction-purification

2.1 Enzymes

(a) Ribonuclease A: Calbiochem (deoxyribonuclease free). This obviates the need to inactivate deoxyribonuclease contaminants by boiling (see ref 29).

(b) T4 polynucleotide kinase: Pharmacia LKB Biotechnology.

(c) T4 DNA ligase: New England Biolabs.

(d) Klenow fragment of *E. coli* DNA polymerase I: Boehringer Mannheim Biochemicals.

(e) Restriction endonucleases: Bethesda Research Laboratories.

Commercial sources for enzymes are provided only as a guide; other suppliers may be superior, and reliable batches of all of these enzymes may be obtained from a single manufacturer.

2.2 Other reagents

(a) 2′-Deoxynucleoside 5′-triphosphates (dATP, dCTP, dGTP, and dTTP) and adenosine 5′-triphosphate (ATP): Pharmacia LKB Biotechnology, supplied as ultrapure 100 mM solutions at pH 7.5. The use of deoxynucleoside triphosphates of the highest possible purity is recommended (see Section 1.2).

(b) Phenol: Bethesda Research Laboratories redistilled nucleic acid grade. Phenol must be buffered prior to use (e.g. using protocol in refs 29 or 30).

2.3 Bacterial and phage strains

(a) JM101 = *E. coli* K12 Δ *(lac-proAB)* *supE thi traD36*/F′ *proA*$^+$*B*$^+$, *lacI*^{Q}, *lacZΔM15* (34).

(b) BMH 71–18 *mutL* = *E. coli* K12 Δ *(lac-proAB)* *supE thi mutL*::Tn10/F′ *proA*$^+$*B*$^+$, *lacI*^{Q}, *lacZΔM15* (16).

(c) M13K07 has the gene II from M13mp1 containing the mutation of G→T at position 6125 (gene II protein Met40→Ile) and insertions of the origin of replication of p15A and the kanamycin resistance gene from Tn903 at the *Ava*I site (position 5825) of M13 (28).

For long-term storage, fresh overnight cultures of these *E. coli* strains should be adjusted to 10–15% (v/v) glycerol and then stored at −70°C. The strains should be restreaked about once a month from the frozen stocks on to minimal glucose plates to maintain the episome which is essential for infection with the M13K07 helper phage.

Escherichia coli JM101 and M13K07 may be obtained from Pharmacia LKB Biotechnology; *E. coli* BMH 71–18 *mutL* can be obtained from the author at the Department of Protein Engineering, Genentech Inc., 460 Point San Bruno Boulevard, South San Francisco, CA 94080, USA. There are many suitable alternative strains for JM101 including XL-1 blue, BMH71–18, TG1, MV1184, and MV1190. XL-1 blue (Stratagene) has the advantage that the episome carries a tetracycline resistance marker (Tn10) which allows it to be readily maintained. This obviates the need for minimal media where bacterial growth is rather slow.

2.4 Media

(a) 2 × TY (2YT): 16 g bacto-tryptone, 10 g yeast extract, 5 g NaCl, add water to 1 litre and adjust pH to 7.4 with NaOH.

(b) TYE plates: 15 g agar, 15 g bacto-tryptone, 10 g yeast extract, 8 g NaCl, add water to 1 litre.

(c) H top agar: 8 g bacto-agar, 10 g bacto-tryptone, 8 g NaCl, add water to 1 litre.

(d) 10 × M9 salts: 70 g $Na_2HPO_4.2H_2O$, 30 g KH_2PO_4, 5 g NaCl, 10 g NH_4Cl, add water to 1 litre.

(e) Minimal glucose plates: 15 g bacto-agar, 888 ml water; after autoclaving add 100 ml 10 × M9 salts, 10 ml 20% (w/v) glucose solution, 1 ml 1 M $MgSO_4$ and 1 ml thiamine (2 mg/ml).

2.5 Preparation of single-stranded phagemid DNA

2.5.1 M13K07 helper phage stock preparation

The procedure in *Protocol 1* gives high titres of M13K07 helper phage ($\geq 10^{11}$ p.f.u./ml) which are required for obtaining high yields of ss phagemid DNA.

M13K07 phage stock may be stored at 4°C and used until template yield (see *Protocol 2*) is significantly diminished (usually 1–2 years). It is prudent (but not normally necessary) to include kanamycin (70 μg/ml) in the media in preparing both M13K07 stocks and phagemid template, in order to maintain the Tn903 (*Kan*[r]) insert in the helper phage (28).

Protocol 1. Preparation of M13K07 helper phage stock

1. Prepare a dilute stock of M13K07 phage by dipping a toothpick into the working stock ($\sim 10^{11}$ p.f.u./ml) and then into 1 ml 2 × TY media. Vortex briefly and dip a toothpick into the diluted phage stock and then into 1 ml 2 × TY media.

2. Take a sterile loop and streak from the most dilute phage stock on to a TYE plate.

3. Overlay the phage with 3 ml H top agar (45°C) containing 30 μl JM101 cells (fresh overnight culture) and incubate overnight at 37°C.

4. Inoculate 5 ml 2 × TY media with 50 μl of a fresh overnight (saturated) culture of JM101 cells.[a]

5. Poke a toothpick into a single well-isolated plaque (from step 3) and use it to infect the culture with M13K07 helper phage.

6. Incubate at 37°C with vigorous aeration for 6 h.

7. Pellet the cells by centrifugation at 6000 g at 4°C for 7 min.

8. Inoculate 200 ml 2 × TY media with 2 ml of a fresh overnight culture of JM101 cells and incubate at 37°C with vigorous aeration until $A_{550} \sim 0.5$.

9. Infect the culture with 5 ml of the phage containing supernatant (from step 7) and incubate at 37°C with vigorous aeration for 6 h.

10. Pellet the cells by centrifugation at 6000 g at 4°C for 15 min.

11. Sterile filter the culture supernatant through a 0.2 μm filter (Nalgene tissue culture type) and store in aliquots at 4°C.

[a] Smaller quantities of M13K07 may be reliably prepared in a single growth step by modification of the protocol above as follows: increase the culture volume at step 4 to 30 ml, incubate at 37°C with vigorous aeration until $A_{550} \sim 0.5$. Proceed as for steps 5 through 7 and then 11 above (i.e. omit steps 8 to 10).

2.5.2 Template preparation

Template is prepared in an identical manner for site-directed mutagenesis and for dideoxy nucleotide sequencing (see *Protocol 2*). Time invested in producing template of good quality is well repaid in terms of the efficiency of mutagenesis and the ease of nucleotide sequencing of putative mutants. Ribonuclease treatment of the template is not essential but is recommended to remove contaminating RNA derived from lysed cells, which can serve as primers for Klenow in the *in vitro* DNA synthesis reaction (12).

The yield of phagemid template is highly dependent upon the phagemid vector, the *E. coli* host strain and the batch of M13K07 used (28). On changing any of these variables it is advisable to check the yield of template (see *Protocol 2*) and to analyse the DNA by agarose gel electrophoresis (see refs 29 and 30 for suitable protocols). The ideal and usual situation is a high yield of phagemid template with barely detectable amounts of ss M13K07 DNA (28). It is also worthwhile to check the yield of template if a large number of mutants are to be made with the same template or as a diagnostic tool if problems are encountered.

Protocol 2. Template preparation for mutagenesis or dideoxy
sequencing

1. Poke a toothpick into a phagemid colony on a fresh minimal glucose plate[a] and use it to inoculate 5 ml 2 × TY media containing an appropriate antibiotic (for selection).

2. Add 50 μl M13K07 helper phage supernatant ($\geq 10^{11}$ p.f.u./ml) and incubate at 37°C for 12–18 h with vigorous aeration.

3. Pellet the cells by centrifugation at 6000 g at 4°C for 7 min.

4. Pour the supernatant into a fresh centrifuge tube, taking care not to disturb the cell pellet.

5. Add 900 μl 20% (w/v) polyethylene glycol (PEG) 6000, 2.5 M NaCl and incubate at 20–25°C (room temperature) for 15–20 min.

6. Pellet the phagemid by centrifugation at 12000 g at 20°C (cold rotor from step 3 is acceptable) for 10 min and pour off the supernatant. There should be a small but clearly visible white pellet. If there is no pellet (or if it is barely visible) then do **not** continue with the template preparation!

7. Remove all traces of PEG (inhibits Klenow) clinging to the walls of the tube by briefly respinning and carefully aspirating the residual supernatant.

8. Resuspend the pellet in 500 μl TE buffer[b] containing 0.1 mg/ml ribonuclease A.

9. Transfer the resuspended phagemid particles to 1.5 ml microcentrifuge tubes using disposable plastic transfer pipettes (do not contaminate pipettmen).

10. Add 1 vol. phenol/CHCl₃ (1:1, v/v) and vortex vigorously for 30 sec. Spin in a microcentrifuge for 5 min to separate the phases and carefully transfer ~400 μl of the upper (aqueous) phase to a fresh tube, taking care not to disturb the debris at the phase interface.

11. Perform a second phenol/CHCl$_3$ extraction (as step 10) and transfer ~300 μl of the aqueous phase into a fresh tube. If there is still a large amount of debris visible it is advisable to perform a third phenol/CHCl$_3$ extraction step.

12. Add 0.5 volume 7.5 M ammonium acetate (made up in glass-distilled water without adjusting the pH) and 3 vol. ethanol and incubate for 10 min at room temperature. (Freezing the samples on dry ice is not necessary because the concentration of DNA is high and may result in less pure DNA preparations.)

13. Spin for 5 min in a microcentrifuge to pellet the DNA and discard the supernatant with a flick of the wrist (pellet is usually not visible).

14. Add 1 ml 70% (v/v) ethanol (room temperature) to remove salt trapped in the DNA pellet. Spin for 3 min in a microcentrifuge and discard the supernatant.

15. Remove traces of ethanol using a SpeedVac evaporator (Savant) for 5 min.

16. Resuspend the pellet in 40 μl TE buffer and store at $-20\,^{\circ}$C until required.

17. **Optional** Check the yield of template by measuring the absorbance of an aliquot at 260 nm in a quartz cuvette containing TE buffer. 1 OD$_{260nm}$ $\equiv$ 37 μg for ss DNA and the average molecular weight per nucleotide is 325 daltons (30). A good yield of template is $\geq$ 4 μg/ml culture but this may be affected by a number of factors (see Section 2.5.2).

---

^{a} Minimal glucose plates are used to select for maintenance of the episome (F') which is essential in order to infect the cells with M13K07 helper phage to obtain template.
^{b} TE buffer is 10 mM Tris–HCl (pH 8.0) 1 mM EDTA.

2.6 Construction of site-directed mutations

A timetable for a typical mutagenesis experiment is shown below:

Day 1 Phosphorylate oligonucleotides (see *Protocol 3*) and initiate muta-enesis reactions (see *Protocol 4*).

Day 2 Transform BMH71–18 *mutL* with heteroduplex DNA (see *Protocol 4*).

Day 3 Prepare minipreps of ds DNA from mutagenesis pools (see *Protocol 4*). Enrich pools for mutant DNA as appropriate.
Restriction-selection and then transformation of JM101 (*Protocol 5*).
Restriction-purification of linearized DNA, self-ligation and transformation of JM101 (*Protocol 6*).

Day 4 Inoculate cultures for template preparation (see *Protocol 2*).

Day 5 Prepare template and nucleotide sequence (see *Protocol 2* and reference 31).

2.6.1 Phosphorylation of oligonucleotides

Mutagenic oligonucleotide primers are phosphorylated prior to use in mutagenesis (see *Protocol 3*) to enable joining of the ends of the nascent DNA strand by T4 DNA ligase.

Protocol 3. Phosphorylation of oligonucleotides

1. To a 1.5 ml microcentrifuge tube add 50 pmol of the mutagenic oligonucleotide,[a] 2 μl 10 × kinase buffer,[b] 1 μl 100 mM DTT and 4 μl 5 mM ATP. Add glass-distilled water to give a final volume of 20 μl.

2. Add 5 units T4 polunucleotide kinase and incubate for 15–30 min at 37°C.

3. Use immediately for mutagenesis reaction or alternatively incubate for 10 min at 70°C (to stop the reaction) and then store at −20°C until required.

[a] The approximate concentration of oligonucleotides may be estimated as described for ss phagemid DNA (see step 17 of *Protocol 2*).
[b] 10 × kinase buffer is 500 mM Tris–HCl (pH 8.0), 100 mM $MgCl_2$.

2.6.2 Construction and transformation of heteroduplex DNA

For *in vitro* construction of heteroduplex DNA (see *Protocol 4*), both template and primer should be of the highest possible purity and a small molar excess of oligonucleotide over template should be used (1.5- to 10-fold). Excessive quantities of mutagenic oligonucleotide may promote priming at additional sites to the target site leading to unwanted mutations. Substochiometric quantities of primer should also be avoided since this will increase the relative amount of wild-type DNA which must be overcome in the mutant enrichment procedures described in Sections 2.6.3 and 2.6.4.

The concentration of both template (see *Protocol 2*) and oligonucleotide (see *Protocol 3*) may be estimated after measuring the absorbance at 260 nm. However, for the simultaneous construction of large numbers of mutants it is relatively time consuming and tedious to measure the concentration of all templates. An acceptable empirical alternative is to use a similar fraction of DNA template preparation which routinely gives a strong clear sequence in dideoxynucleotide sequencing.

Protocol 4. Mutagenesis and transformation of heteroduplex DNA

T7 DNA polymerase is used in this *Protocol* in place of Klenow because it permits the reaction time to be reduced from overnight to 1 h and results in improved mutant yields (see Section 1.2 and refs 42 and 43).

1. To a 1.5 ml microcentrifuge tube add 1 μl (2.5 pmol) phosphorylated muta-genic primer (see *Protocol 3*), 1–3 μl (~0.5 μg/μl) template (see *Protocol 2*), 2 μl 5 × SEQ buffer,[a] and glass-distilled water to a total volume of 10 μl.

2. Place the sample tube into a small beaker of boiling water and then immediately allow it to cool to room temperature (~20 min).

3. Spin the sample for a few seconds in a microcentrifuge to recover con-densation from the lid of the tube and place on ice.

4. Add 2 μl 5 × SEQ buffer,[a] 1 μl 5 mM ATP, 1 μl 5 mM dNTPs,[b] 1 μl 100 mM DTT, 2 μl 1 μg/μl acetylated bovine serum albumin (United States Biochemicals) and glass-distilled water to give a final volume of 20 μl (after the enzyme additions in step 5).

5. Add 3 units T4 DNA ligase, 5 units T4 polynucleotide kinase, and 0.6 units T7 DNA polymerase (United States Biochemicals) and incubate at 37°C for 1 h.

6. Transform 200 μl CaCl$_2$-treated (competent) BMH 71–18 *mutL* cells with 1–5 μl of the mutagenesis reaction. Competent cells may be prepared as described in ref. 25, and give good results used fresh or after flash freezing in the presence of 10% (v/v) glycerol and storage at −70°C. (High efficiency competent cells prepared as in refs 40 and 41 may be used; however, this is not necessary with the large amounts of heteroduplex DNA constructed.)

7. Add 1 ml 2 × TY media to the transformed cells and incubate at 37°C with minimal shaking for 30–60 min.

8. Add the transformed cells to 25 ml 2 × TY media containing an appropriate antibiotic to select for transformed cells and incubate at 37°C overnight with vigorous aeration. This level of dilution is recommended to minimize any problems associated with debris from untransformed cells.

9. Prepare a miniprep of ds phagemid DNA from 5 ml of the culture and resuspend the miniprep DNA in 50 μl TE buffer (see *Protocol 2*), e.g. use an alkaline lysis procedure to prepare miniprep DNA as described in refs 29 and 30.

[a] 5 × SEQ buffer is 200 mM Tris–HCl (pH 7.5), 80 mM MgCl$_2$.
[b] 5 mM dNTPs is 5 mM dATP, 5 mM dTTP, 5 mM dCTP, and 5 mM dGTP and is prepared by diluting 100 mM dNTP stocks (pH 7.5) into TE buffer.

2.6.3 Enrichment for mutant DNA by restriction-selection

The amount of ds phagemid DNA used in the restriction-selection digest (≤100 ng; see *Protocol 5*) is kept as small as possible and a large amount (10–20 units) of the necessary restriction enzyme is used in order to maximize the fraction of wild-type DNA in the pool which is cleaved and thus minimize the background of wild-type progeny. A unit of restriction enzyme is commonly defined as the amount of enzyme which is required for complete cleavage of $1\,\mu$g of bacteriophage λ DNA in one hour at $37\,^{\circ}$C in a total reaction volume of $50\,\mu$l. Thus the conditions in *Protocol 5* nominally represent a $\sim$100- to 200-fold excess digest. This is not wasteful since effective cleavage of wild-type DNA is essential to obtain good mutant yields and several factors may reduce the rate of cleavage at the target site. First, the batch of enzyme used may have less activity than indicated on the tube or its activity may be lowered by contaminants in the DNA preparation. Second, restriction enzyme cleavage rates are dependent upon the flanking sequences and whether the substrate is linear or supercoiled. For enzymes including *Sal*I and *Sca*I at least 10-fold more enzyme was required to completely digest pBR322 plasmid DNA as compared to λ DNA (see New England Biolabs, 1988–9 catalogue). These digestion conditions should (and generally do) ensure that none of the clones which are sequenced are wild-type. However, 10–25% of clones sequenced are commonly non-target mutants (see Section 3.1).

Protocol 5. Restriction-selection[a] to enhance mutant yield

1. Digest $1\,\mu$l ($\leq$100 ng) of miniprep DNA (from step 9 of *Protocol 4*) with 20 units of the restriction endonuclease corresponding to the site **removed** by the mutagenic oligonucleotide (1–2 h under the buffer and salt conditions and at the temperature recommended by the manufacturer).

2. Transform CaCl$_2$-treated competent JM101 cells (see steps 6 and 7 of *Protocol 4*) directly with 1–20% of the digest from step 1.

3. Plate out on to minimal glucose plates[b] containing an appropriate antibiotic to select for the transformed cells and incubate overnight at $37\,^{\circ}$C.

4. Inoculate 2–5 cultures (per mutagenesis reaction) for template preparation as described in *Protocol 2*. Prepare ss DNA from only 2 clones per mutagenesis reaction and analyse the nucleotide sequence by Sanger's dideoxynucleotide method (e.g. as described in ref. 31). For the remaining clones (if any) the phage containing culture supernatant (step 4 of *Protocol 2*) may be stored at $4\,^{\circ}$C for at least a few days without sterile filtering in case they are required.

5. If ds DNA is needed, e.g. for subcloning mutant fragments, then the following alternative strategies are recommended:

    (**A**) At step 4 above, inoculate 2 cultures for each clone and add M13K07 helper phage to one. Freeze the cell pellets corresponding to the non-M13K07 infected cultures at $-70\,^{\circ}$C. After nucleotide sequence verification of mutants, prepare miniprep DNA from corresponding cell pellets as required.

    (**B**) At step 4 above, streak out the candidate mutants on to selective plates. Further cultures can then be grown to prepare ds or ss DNA.

(**C**) Prepare ds DNA *in vitro* by primer-directed extension of the sequence-verified ss DNA, e.g. as described in ref. 30.

[a] Unique restriction site is **removed** by the mutagenic oligonucleotide (see *Figure 2*).
[b] See footnote [b] to *Protocol 2*.

2.6.4 Enrichment for mutant DNA by restriction-purification

The amount of DNA ($\sim 1\,\mu$g) used in the restriction-purification digest (see *Protocol 6*) is almost invariably sufficient that DNA linearized at the new restriction site shows up as as a clearly visible ethidium bromide-stained band in a 5% polyacrylamide gel. Above this band is often a broad smear which is probably due to fragments of chromosomal DNA in the miniprep. Undigested (wild-type) DNA does not efficiently enter the gel. It is exceedingly rare to sequence any wild-type clones obtained after electroelution and self-ligation of the linearized DNA. However, 10–25% of clones sequenced are commonly non-target mutants (see Section 3.1).

Protocol 6. Restriction-purification [a] of mutated DNA

1. Digest $10\,\mu$l ($\sim 1\,\mu$g) of the miniprep DNA with the restriction endonuclease corresponding to the unique site **introduced** by the mutagenic oligonucleotide. This is normally plenty of DNA to recover the linearized mutated DNA, even if the mutant yield is very low (a few per cent).

2. Electrophorese the DNA sample on a native 5% polyacrylamide gel (e.g. 200 × 200 × 1.5 mm) containing 1 × TBE buffer [b] for 2 h at 200 V (constant). Detailed protocols for the analysis of DNA fragments by polyacrylamide gel electrophoresis may be found in refs 29 and 30. (It is not generally necessary to purify the digested DNA by phenol/CHCl₃ extraction and ethanol precipitation (see step 9 below) in order to electrophorese the sample reliably.)

3. Stain the gel after electrophoresis by gently shaking in 1 × TBE (running) buffer containing $\sim 2 \times 10^{-4}\%$ (w/v) ethidium bromide (i.e. $2\,\mu$g/ml) for ≥ 10 min at room temperature.

4. View the gel by transillumination with long wave ($\lambda = 300$–360 nm) UV light. Excise a gel slice containing the linearized DNA. (Do **not** use short-wave UV light since this will damage the DNA, e.g. by photodimerization of thymidine).

5. Place the gel slice into a dialysis bag (prepared as in ref. 29) containing 200–400 μl 1 × TBE buffer and **no bubbles**.

6. Electroelute the linearized DNA, e.g. use 1 × TBE buffer in a mini-gel apparatus cooled in an ice bath (200 V, 30–60 min). The time taken for elution varies greatly with fragment size, applied voltage, buffer, and

Protocol 6. continued

   device used for electroelution. It is therefore advisable to verify under
   long-wave UV that the ethidium bromide-stained DNA has indeed been
   electroeluted.

7. Reverse the polarity of the current for 30 sec to release the DNA from
   the walls of the dialysis tubing.

8. Carefully remove the contents of the dialysis bag, rinse the bag with a
   further $200\,\mu l$ 1 × TBE buffer and pool these samples.

9. Extract the digested DNA with 1 vol. phenol/CHCl$_3$ (1:1, v/v) as for
   Protocol 2, step 11 (one extraction only). Add 0.1 vol. 3 M sodium
   acetate (pH 4.5–5.5), 2.5 vol. of ethanol and place on dry ice until
   frozen. (The concentration of DNA is low and the dry-ice step is require
   to precipitate the DNA in high yield.) Pellet the DNA, rinse with 70%
   (v/v) ethanol, dry the pellet and resuspend as in *Protocol 2*, steps 13–16.

10. Self-ligate the linearized DNA: to a 1.5 ml microcentrifuge tube add
    $10\,\mu l$ electroeluted DNA (20–50% of total), $2\,\mu l$ 10 × LB buffer,^{c} $1\,\mu l$
    5 mM ATP, $1\,\mu l$ 0.1 M DTT, 5–10 units of T4 DNA ligase and adjust the
    final volume to $20\,\mu l$ with glass-distilled water.

11. Incubate the ligation reaction for 30 min to overnight at 14 °C. 30 min is
    adequate time for self-ligation of linear DNA wth staggered ('sticky')
    ends using high efficiency competent cells (prepared as in refs 40 or 41).
    Overnight incubation is recommended for a blunt-ended ligation or
    where low efficiency (CaCl$_2$-treated) competent cells are used (prepared
    as in ref. 25).

12. Transform JM101 cells with 5–50% of the ligation reaction (see *Protocol
    4*, steps 6–7) and then continue as for *Protocol 5*, steps 3–5.

^{a} Unique restriction site is **introduced** by mutagenic oligonucleotide (see *Figure 2*).
^{b} TBE buffer is 90 mM Tris base, 89 mM boric acid (pH 8.3), 2.5 mM EDTA.
^{c} 10 × LB buffer is 500 mM Tris–HCl (pH 7.5), 100 mM MgCl$_2$.

Linearized DNA may be very rapidly and reliably (but not quantitatively)
recovered from a polyacrylamide gel slice by the following modifications of
Protocol 6: at step 4 place the excised gel slice in a 0.22 μm centrifugal filtration
unit (Millipore catalogue #UFC3 OGV 00) and spin in a microcentrifuge for 10
min. The linearized DNA in the filtrate is then religated without further
treatment as described in step 11 and used to transform high efficiency
competent cells prepared as in ref. 40.

2.6.5 Long-term storage of mutant clones

For each mutant phagemid clone save some of the template preparation used
for nucleotide sequencing (e.g. for further rounds of site-directed mutagenesis).

It is also advisable to prepare a glycerol stock of each mutant clone in the host strain required for expression (see *Protocol 7*) for indefinite storage at $-70\,°C$. Mutants can then be recovered at any time by simply restreaking from the glycerol stock **without** thawing it.

Protocol 7. Preparing mutant clones for long-term storage

1. Poke a toothpick into a phagemid colony (mutant clone in the host strain required for expression) and use it to inoculate 5 ml 2 × TY media containing an appropriate antibiotic for selection.
2. Incubate at $37\,°C$ for 12–18 h with vigorous aeration.
3. Adjust the culture to 10–15% (v/v) glycerol and aliquot 1 ml into a sterile microcentrifuge tube.
4. Flash freeze the culture in a dry-ice/ethanol bath or with liquid nitrogen and store at $-70\,°C$.
5. Regenerate mutant clones as required by scraping the surface of the frozen glycerol stock with a sterile loop and restreaking on to a plate containing an appropriate antibiotic to select for the phagemid. The gycerol stock culture should be kept on dry ice and **not** allowed to thaw out.

---

3. Trouble-shooting procedures

The guiding principle in the construction and verification of mutations is to minimize the total operator time required. If several mutants are being constructed simultaneously I recommend sequencing no more than two clones per mutant (see *Protocol 5*), since this is normally sufficient to obtain the majority of mutants. Sequence no more than a few additional clones (~4) for mutants which are not obtained in the first sweep. Generally, $\geq 90\%$ of target mutants have been obtained by this stage and the trouble-shooting procedures below should be directed against the remaining problematic mutants.

3.1 Likely origin of clones other than target mutant

The nucleotide sequence of clones which do not contain the desired mutation may be categorized into one of several different classes which provides useful clues as to the likely problem. The probable origin of non-target mutant clones is discussed below:

(a) Wild-type clones are occasionally found after restriction-selection (extremely rarely after restriction-purification) which usually reflects failed or inefficient digestion. Another possibility is that the yield of mutants is exceedingly low (0–1%). The latter explanation probably reflects prob-

lems with the particular mutagenic oligonucleotide if other parallel mutagenesis experiments were successful, or a more systematic problem if
many other mutants were not obtained (see Section 3.2).

(b) Sometimes the restriction site mutation is found without the desired silent
mutation ('decoupling'). There are a number of potential explanations
for decoupling including inefficient *in vitro* DNA synthesis followed by
$5'\rightarrow3'$ exonuclease activity *in vivo* or displacement of the mutagenic
oligonucleotide by Klenow after extending all-the-way-around (see Section
1.2). It is prudent to check that the repair deficient host strain is tetracycline resistant since these phenotypes should be linked (*mutL*::Tn10).
Independent *dam*-directed mismatch repair is not a likely explanation for
decoupling since the repair deficient host strain is rather stable if properly
maintained as described in Section 2.3. There are however other DNA
repair systems in *E. coli* which may be responsible (see ref. 15 and
literature cited therein).

(c) Mutant sequences are sometimes found which are very close to the
expected change but contain an additional change(s) (point mutation(s)
or small insertion or deletion) within the region defined by the oligonucleotide. This class of mutants almost certainly reflects problems with
the mutagenic oligonucleotide itself and may necessitate synthesis of a
new primer (see Sections 3.3 and 3.4).

(d) Mutants are sometimes found in which one end of the oligonucleotide has
apparently annealed at the target site whilst the other end has hybridized
spuriously at some additional site to which it shows some homology. This
can generate deletions which may include the site used for restriction-
selection. This problem is most commonly encountered using long oligonucleotides to construct large deletions (35, 36) or a complex mixture of
mutations. However, judicious choice of restriction site in designing the
primer for restriction-selection or restriction-purification (see Section
1.5) and varying the conditions for annealing primer and template (see
Section 3.3) can substantially reduce or even eliminate this problem.

(e) A final class of non-target mutant sequences is where a likely explanation
as to their origin (other than by divine intervention) is not readily forth-
coming.

Failure of a clone to give any detectable sequence may be due to deletion of
the priming site for the sequencing primer [see point (d) above]. Alternatively
there may be limiting amounts of template or systematic problems with
dideoxynucleotide sequencing.

3.2 Checks on the efficiency of mutagenesis

One of the attractive features of mutagenesis using restriction-purification is
that the relative amounts of linearized and uncut DNA provides a crude but

nevertheless useful indicator of the efficiency of mutagenesis. Thus a previously used template/primer combination may be used as a simple diagnostic tool to see whether the yield of mutants has significantly changed.

An alternative and quantitative test for mutant yield provided with some commercial mutagenesis kits is to use a test primer and template which generates an easily detectable phenotypic change (22). For example, Messing's pUC118 and pUC119 phagemids (28) (and also M13mp vectors) contain the *lacZ'* gene which may be induced by isopropyl-β-D-thiogalactoside (IPTG) to express the *lacZ* α-peptide. The α-peptide complements a deletion in widely available *E. coli* strains, such as JM101, to give a functional β-galactosidase. The chromogenic substrate analogue, 5-bromo-4-chloro-3-indolyl β-D-galactoside (known as BCIG or 'Xgal'), is hydrolysed by β-galactosidase to a blue dye which stains the colonies. A mutagenic oligonucleotide which introduces a stop codon into *lacZ'* will prevent α-complementation and result in 'white' (non-blue) colonies. The relative number of blue and white colonies is a reliable guide to the yield of oligonucleotide-directed mutants, as the fraction of white colonies which arise by other events is very small. A drawback of this approach is that a different template and primer are required than for the silent mutation of interest.

3.3 Specificity of priming of mutagenic oligonucleotide

The specificity of priming of the mutagenic oligonucleotide may be checked by using it in a dideoxynucleotide sequencing reaction or in a primer-directed extension reaction followed by restriction digestion as described by Zoller and Smith (37). It is worthwhile to conduct these specificity tests at different ratios of primer to template as it is sometimes possible to reduce or eliminate spurious priming by using a smaller excess of primer. It may also be helpful to include a previously used sequencing primer in these control experiments to evaluate the template and polymerase used in the mutagenesis reaction.

3.4 Purity of mutagenic oligonucleotide

Problems in oligonucleotide synthesis will very often manifest themselves during their subsequent purification by HPLC or by polyacrylamide gel electrophoresis (e.g. large amounts of failure products or low yield of desired oligonucleotide). If the original purification data are not available then the purity of the primer should be checked by polyacrylamide gel electrophoresis after phosphorylation with [γ-^{32}P]ATP (25). The oligonucleotide should give only one major band by autoradiography and may show a few minor contaminants. This provides only a rough guide to the oligonucleotide length since it is not uncommon for oligonucleotides which differ by one or even two bases in length to co-migrate (depending upon base composition). If a ladder of bands is observed then this may represent failure sequences in the oligonucleotide preparation which may result in unwanted base changes or nucleotide deletions during mutagenesis (see Section 3.1). An alternative explanation is that the T4 polynucleotide kinase is contaminated with exonuclease.

3.5 Extent of phosphorylation of mutagenic oligonucleotide

Quantitative phosphorylation of the mutagenic primer is essential for obtaining high yields of mutants. The extent of phosphorylation (usually ~100%) may be checked by electrophoresing the oligonucleotide on a polyacrylamide/8 M urea gel and detecting the oligonucleotide by UV shadowing (see refs 30, 32, 33, for detailed protocols). The oligonucleotide migrates faster after phosphorylation by virtue of the extra negative charge allowing the extent of phosphorylation to be assessed.

4. Advantages and limitations of restriction-selection and restriction-purification

Mutagenesis using restriction-selection and restriction-purification are useful tools in protein engineering where the necessary restriction site change in a gene can be made silently—without introducing an amino acid change into the corresponding protein. However, these techniques are less useful in structure–function analysis of non-coding sequences such as expression control elements where other mutagenesis methods (reviewed in refs 1–3) may be more appropriate. Restriction-selection is attractive because it is exceedingly simple and reliable and the total operator time required is low. Furthermore, selection against wild-type progeny is very tight, which enables mutants to be recovered easily, even if the mutagenesis frequency is low or variable (e.g. where the mutagenic oligonucleotide primes inefficiently at the target site on the template or where reagents are suboptimal). Restriction-purification involves slightly more effort than restriction-selection. However it has the advantage that purification of linearized DNA from a gel provides a useful qualitative check on the efficiency of mutagenesis (see Section 3.2). It is often desirable to combine mutants by ligation of restriction fragments containing different mutations. The presence of a unique restriction site in the vicinity of the silent mutation of interest provides a very useful tool for verifying clones generated, which is much less time-consuming than having to re-sequence each mutant locus. New restriction sites installed during mutagenesis by restriction-purification can be used for restriction-selection in constructing further mutations (24).

Transformation of heteroduplex DNA into a repair deficient host strain increases the yield of mutants by up to 10-fold (12) and ensures that the restriction site change and the mutation of interest are tightly coupled, which is essential to the success of the restriction-selection and restriction-purification procedures. The frequency of mutants before enrichment is generally 5–40%, which is almost invariably sufficient for the linearized DNA in the restriction-purification procedure to be readily detected and recovered. The use of repair deficient host strains also improves the reliability of restriction-selection since it reduces the background of wild-type DNA molecules which must be eliminated.

As an example of the efficiency and reliability of restriction-selection using repair deficient host strains, 64 single amino acid replacements were made in human growth hormone (hGH) and a total of only ~150 clones were actually sequenced (38). This large mutagenesis project was greatly facilitated by the construction of a synthetic gene for hGH containing unique restriction sites about every 15 codons (39). This enabled mutagenesis both by restriction-selection and by cloning cassettes and thus permitted great flexibility in mutagenesis strategy.

A problem which is intrinsic to oligonucleotide-directed mutagenesis by heteroduplex construction is that of mispriming events (see Section 1.2). However, this is usually only a significant difficulty in using long oligonucleotides to generate complex mutations. In contrast to other widely used mutagenesis methods using mismatched oligonucleotides, the restriction-selection and restriction-purification procedures can be designed to substantially reduce or even eliminate this problem and have enabled complex mutations such as that in *Figure 3* to be made in a single step. Thus this extends the scope of mutagenesis using mismatched oligonucleotides by increasing the complexity of mutants which can be readily constructed.

Acknowledgements

The author gratefully acknowledges the contribution of many former and present colleagues to these methods including Dr Jim Wells of the Department of Protein Engineering at Genentech and Dr Greg Winter of the Laboratory of Molecular Biology, Hills Road, Cambridge, CB2 2QH, UK. Dr Steve Bass, Dr Ron Greene, Brian Cunningham and Mark Dennis are thanked for helpful comments on the manuscript and Wayne Anstine is thanked for MacWizardry in preparing the figures.

References

1. Smith, M. (1985). *Annual Reviews in Genetics*, **19**, 423.
2. Carter, P. (1986). *Biochemical Journal*, **237**, 1.
3. Kunkel, T. A. (1988). In *Nucleic Acids and Molecular Biology* (ed. F. Eckstein and D. M. J. Lilley), Vol. 2, p124, Springer-Verlag, Berlin and Heidelberg.
4. Wells, J. A., Vasser, M., and Powers, D. B. (1985), *Gene*, **34**, 315.
5. Engels, J. W. and Uhlmann, E. (1989). *Angewandte Chemie, International Edition in English*, **28**, 716.
6. Saiki, R. K., Scharf, S., Faloona, F., Mullis, K. B., Horn, G. T., Erlich, H. A., and Arnheim, N. (1985). *Science*, **230**, 1350.

7. Higuchi, R., Krummel, B., and Saiki, R. K. (1988). *Nucleic Acids Research*, **16,** 7351.
8. Vallette, F., Mege, E., Reiss, A., and Adesnik, M. (1989). *Nucleic Acids Research*, **17,** 723.
9. Craik, C. S., Largman, C., Fletcher, T., Roczniak, S., Barr, P. J., Fletterick, R., and Rutter, W. J. (1985). *Science*, **228,** 291.
10. Kramer, W., Ohmayer, A., and Fritz, H.-J. (1988). *Nucleic Acids Research*, **16,** 7207.
11. Venkitaraman, A. R. (1989). *Nucleic Acids Research*, **17,** 3314.
12. Carter, P., Bedouelle, H., and Winter, G. (1985). *Nucleic Acids Research*, **13,** 4431.
13. Kunkel, T. A. (1985). *Proceedings of the National Academy of Sciences of the USA*, **82,** 488.
14. Tabor, S. and Richardson, C. C. (1987). *Proceedings of the National Academy of Sciences of the USA*, **84,** 4767.
15. Lahue, R. S., Au, K. G., and Modrich, P. (1989). *Science*, **245,** 160.
16. Kramer, B., Kramer, W., and Fritz, H.-J. (1984). *Cell*, **38,** 879.
17. Laengle-Rouault, F., Maenhaut-Michel, G., and Radman, M. (1986). *EMBO Journal*, **5,** 2209.
18. Enea, V., Vovis, G. F., and Zinder, N. D. (1975). *Journal of Molecular Biology*, **96,** 495.
19. Carter, P. (1985). PhD Thesis, p. 125. University of Cambridge, UK.
20. Taylor, J. W., Ott, J., and Eckstein, F. (1985). *Nucleic Acids Research*, **13,** 8765.
21. Vandeyar, M. A., Weiner, M. P., Hutton, C. J., and Batt, C. A. (1988). *Gene*, **65,** 129.
22. Kramer, W., Drutsa, V., Jansen, H.-W., Kramer, B., Pflugfelder, M., and Fritz, H.-J. (1984). *Nucleic Acids Research*, **12,** 9441.
23. Stanssens, P., Opsomer, C., McKeown, Y. M., Kramer, W., Zabeau, M., and Fritz, H.-J. (1989) *Nucleic Acids Research*, **17,** 4441.
24. Wells, J. A., Cunningham, B. C., Graycar, T. P., and Estell, D. A. (1986). *Philosophical Transactions of the Royal Society of London*, *A*, **317,** 415.
25. Carter, P. (1987). In *Methods in Enzymology* (ed. R. Wu), Vol. 154, p. 382. Academic Press, London.
26. Dente, L., Cesareni, G., and Cortese, R. (1983). *Nucleic Acids Research*, **11,** 1645.
27. Dotto, G. P., Enea, V., and Zinder, N. D. (1981). *Virology*, **114,** 463.
28. Vieira, J. and Messing, J. (1987). In *Methods in Enzymology* (ed. R. Wu), Vol. 153, p. 3. Academic Press, London.
29. Sambrook, J., Fritsch, E. F. and Maniatis, T. (1989). *Molecular Cloning. A Laboratory Manual* 2nd edn. Cold Spring Harbor Laboratory, Cold Spring Harbor, New York.
30. Ausubel, F. M., Brent, R., Kingston, R. E., Moore, D. D., Seidman, J. G., Smith, J. A., and Struhl, K. (ed.) (1987, 1988, 1989). *Current Protocols in Molecular Biology*. John Wiley, New York.
31. Bankier, A. T., Weston, K. M., and Barrell, B. G. (1987). In *Methods in Enzymology* (ed. R. Wu), Vol. 155, p. 51. Academic Press, London.
32. Gait, M. J. (ed.) (1984) *Oligonucleotide Synthesis—A Practical Approach*. IRL Press, Oxford.

33. Carter, P., Bedouelle, H., Waye, M. M. Y., and Winter, G. (1985). In *Oligonucleotide Site-Directed Mutagenesis in M13*. Anglian Biotechnology Ltd, Colchester.
34. Messing, J. (1979). *Recombinant DNA Technical Bulletin*, **2**, 43.
35. Osinga, K. A., Bliek, A. M. Van der, Horst, G. Van der, Groot Koerkamp, M. J. A., Tabak, H. F., Veeneman, G. H., and Boom, J. H. Van (1983). *Nucleic Acids Research*, **11**, 8595.
36. Chan, V.-L. and Smith, M. (1984). *Nucleic Acids Research*, **12**, 2407.
37. Zoller, M. J. and Smith, M. (1987). In *Methods in Enzymology* (ed. R. Wu), Vol. 154, p. 329. Academic Press, London.
38. Cunningham, B. C. and Wells, J. A. (1989). *Science*, **244**, 1081.
39. Cunningham, B. C., Jhurani, P., Ng, P., and Wells, J. A. (1989). *Science*, **243**, 1330.
40. Chung, C. T. and Miller, R. H. (1988). *Nucleic Acids Research*, **16**, 3580.
41. Hanahan, D. (1983). *Journal of Molecular Biology*, **166**, 557.
42. Bebenek, K. and Kunkel, T. A. (1989). *Nucleic Acids Research*, **17**, 5408.
43. Tabor, S. and Richardson, C. C. (1989). *Journal of Biological Chemistry*, **264**, 6447.

2

Site-directed *in vitro* mutagenesis using uracil-containing DNA and phagemid vectors

PATRICIA D. YUCKENBERG, FRANK WITNEY,
JANET GEISSELSODER, and JOHN McCLARY

1. Introduction

Oligonucleotide directed *in vitro* mutagenesis is a widely used procedure for the study of the structure and function of DNA and the protein for which it codes, for studies on regulation of gene expression and DNA replication, and for modification of food and therapeutic proteins. A wide variety of techniques is available for performing *in vitro* mutagenesis (reviewed in ref. 1 and this book). One strategy is to clone the segment of DNA to be mutated into a vector whose DNA exists in both single- and double-stranded forms. An oligonucleotide complementary to the region to be altered, except for a limited internal mismatch, is hybridized to a single-strand copy of the DNA. A complementary strand is then synthesized by DNA polymerase using the oligonucleotide as primer. Ligase is used to seal the new strand to the 5'-end of the oligonucleotide. The double-stranded DNA, homologous except for the intended mutation, is then transformed into *E. coli*, resulting in two classes of progeny; the parental and those carrying the oligonucleotide-directed mutation. Since there are both parental and mutant progeny, no more than half of the progeny will be mutant. In practice, a much lower fraction is usually obtained. Recently, a number of methods have been developed which are more efficient. These methods create an asymmetry between the two strands of the heteroduplex and permit selection against the wild-type strand (see also Chapters 1 and 3).

One of these methods, originally described by Kunkel (2, 3), provides a very strong selection against the non-mutagenized strand of a double-stranded DNA. The approach is outlined in *Figure 1*. When DNA is synthesized in a *dut ung* double mutant bacterium, the nascent DNA carries a number of uracils in thymine positions as a result of the *dut* mutation which inactivates the enzyme dUTPase and results in high intracellular levels of dUTP. The *ung* mutation inactivates uracil N-glycosylase, thus allowing the

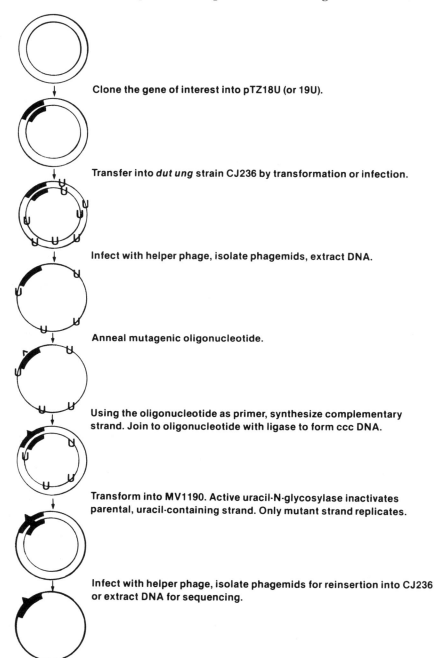

Figure 1. Steps involved in *in vitro* mutagenesis using uracil-containing DNA and phagemid vectors. (Reprinted with permission of Eaton Publishing.)

incorporated uracil to remain in the DNA. This uracil-containing strand is then used as the template for the synthesis *in vitro* of a complementary strand primed by an oligonucleotide containing the desired mutation. When the resulting double-stranded DNA is transformed into a cell with a proficient uracil N-glycosylase, the uracil-containing strand is inactivated with high efficiency, leaving the non-uracil-containing survivor to replicate. Typical mutagenesis frequencies obtained with this method are greater than 50%, a rate high enough to allow identification of mutants by direct sequence analysis.

This method was originally described for use with M13 vectors (2, 3, 4). In recent years, a different type of vector useful for *in vitro* mutagenesis has been developed. The origin of replication from filamentous single-stranded phage (i.e. M13, fd, f1) has been inserted into small, high copy number plasmids. These vectors, termed phagemids (phage plus plasmid), replicate as double-stranded plasmids until the host is superinfected with a helper phage. The proteins coded by the helper phage act at the origin of replication to permit the phagemid DNA to be replicated, packaged in phage protein, and extruded from the cell as if it were a single-stranded phage. Phagemid vectors combine the advantages of plasmids (high-copy number, stability of cloned inserts) with the features of phage (ability to produce single-stranded DNA).

Although a wide variety of phagemid vectors may be used for mutagenesis with uracil-containing DNA, we are most familiar with the vectors pTZ18U and pTZ19U (5, *Figure 2*). pTZ18U and pTZ19U (6) are derivatives of the pUC plasmids constructed in the Messing laboratory (7). In addition to colE1 replication functions and the gene coding for ampicillin resistance, they carry the *lac* operator and promoter and the first 145 codons of the β-galactosidase gene (*lacZ'* fragment). This α-peptide of β-galactosidase can complement a truncated host β-galactosidase fragment (ΔM15 deletion) made by the bacterial host strain and results in a functional enzyme. Colonies of the appropriate *Escherichia coli* host carrying the plasmid will be blue in the presence of the inducer isopropyl-β-D-thiogalactopyranoside (IPTG) and the substrate 5-bromo-4-chloro-3-indolyl-β-D-galactopyranoside (X-gal). In addition, a multiple cloning site (called a polylinker) which carries numerous unique restriction enzyme recognition sites was incorporated near the beginning of the *lac* DNA. This polylinker contains unique restriction sites for *Eco*RI, *Sst*I, *Kpn*I, *Xma*I, *Sma*I, *Bam*HI, *Xba*I, *Sal*I, *Acc*I, *Hinc*II, *Pst*I, *Sph*I, and *Hind*III. Insertion of DNA into these sites disrupts the DNA coding for the β-galactosidase fragment and results in colourless (or pale blue) colonies when plated in the presence of IPTG and X-gal. The insertion of the f1 origin of replication permits, by superinfection with a helper phage, the formation of phage-like particles containing single-stranded DNA. The plus (or coding) strand of pTZ18U and pTZ19U is packaged. A portion of the sequence of the plus strand is shown in *Figure 2*. In addition, a strong promoter for T7 RNA polymerase has been inserted into the *lac* region without interrupting the

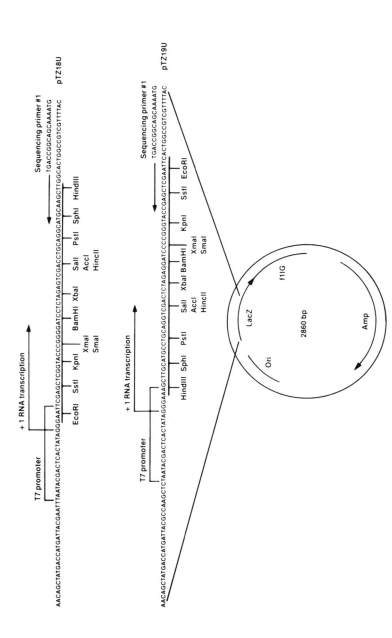

Figure 2. Diagrammatic representation of the features of the pTZ18U and pTZ19U plasmids. The nucleotide sequence of the multiple cloning region of the plasmids are shown and the restriction sites present in the region noted. Features of the plasmid, as discussed in the text, include the β-lactamase gene which confers ampicillin (amp) resistance to *E. coli* containing the plasmid, the *lacZ'* fragment of the β-galactosidase gene, the T7 RNA polymerase promoter region, and the f1 intergenic (IG) region. The latter feature makes it possible to generate single-stranded DNA from pTZ plasmids by superinfection of plasmid-containing strains with appropriate helper phages. As shown, single-stranded DNA generated from pTZ18U or pTZ19U may be sequenced by the dideoxy method, using the universal sequencing primer.

coding sequence. This promoter allows the plasmid DNA to be used to generate high levels of insert-specific RNA by *in vitro* transcription.

2. Mutagenesis procedures

2.1 Bacteriology

Escherichia coli CJ236 carries the following markers: *dut*-1, *ung*-1, *thi*-1, *rel* A-1; pCJ105(Cm^r). The *dut* and *ung* phenotypes are non-reverting and result in occasional uracils substituted for thymine in all DNA synthesized in the bacterium (3). CJ236 contains a functional *lac* operon. All colonies will be blue in the presence of IPTG and X-gal, so it is not useful to include these indicators in the plates.

Escherichia coli MV1190 is Δ(*lac-pro*AB), *thi*, *sup*E, Δ(*srl-rec*A) 306::Tn10(tet^r)[F′:*tra*D36, *pro*AB, *lac*I^q ZΔM15]. In this strain DNA coding for lactose utilization and proline biosynthesis has been deleted from the chromosome. The F′ plasmid carries the proline synthesis genes and a truncated β-galactosidase gene which produces a protein fragment that can be complemented by the α-peptide coded for by the pTZ phagemid. Therefore, cells carrying phagemids with an intact α-peptide-coding sequence will produce blue colonies in the presence of the inducer IPTG and the indicator dye X-gal. Cells carrying phagemids with DNA inserted into this sequence will not produce a functional α-peptide [or produce a poorly functional fragment (8)] and, hence, will generate colourless colonies when IPTG and X-gal are included in the plates.

Both the phagemid particles and the helper phage require pili for entry into the cell. These pili are part of the fertility function of *E. coli*. **Since pili are not formed below 35 °C, all growth should take place at 37 °C.** Both the cloning/ selecting strain MV1190 and the uracil-inserting strain CJ236 contain plasmids, called F′ plasmids, which code for the functions required to manufacture pili. Since the cell must exert energy to maintain these plasmids, they will be lost if selective pressure for their presence is not maintained.

E. coli MV1190 has lost some of the chromosomal genes for synthesis of proline and the entire *lac* operon. The F′ carries the *lac* operon except for a small segment of the β-galactosidase gene and the missing genes for proline synthesis. By requiring synthesis of proline, selection is maintained for the F′. For this reason, MV1190 should be streaked on a glucose-minimal medium plate and well-isolated colonies used to inoculate liquid cultures. The defective β-galactosidase gene can be complemented by the *lac*Z′ segment coded for on the pTZ phagemids, thus allowing blue-white screening for inserts.

The F′ in *E. coli* CJ236 carries the genes for chloramphenicol resistance; therefore in order to maintain the F′, CJ236 should always be grown in the presence of chloramphenicol. A working culture is obtained by streaking from the glycerol stock supplied on to an LB (or H) plate containing chloramphenicol. Well-isolated single colonies from this plate are used as inocula to start **every** overnight culture. These plates can be used for two to three months, then a new plate is made by streaking from the original stock **not** from the plate. The concentration of chloramphenicol **should not exceed** the recommended 30 μg/ml in either liquid or solid media. CJ236 has a doubling time about twice that of MV1190.

2.1.1 Media and antibiotics

For detailed instructions on preparation of media, sterile techniques, titreing of phage, etc., three manuals from Cold Spring Harbor Laboratory are very useful (9, 10, 21). Recipes for media and antibiotic solutions are listed below.

i Media
(a) **LB (L broth)**
  10 g bacto-tryptone, 5 g yeast extract, 5 g NaCl, deionized H_2O to 1 litre.
(b) **2 × YT**
  16 g bacto-tryptone, 10 g yeast extract, 5 g NaCl, deionized H_2O to 1 litre.
(c) **Glucose-minimal media**
  6 g Na_2HPO_4, 3 g KH_2PO_4, 0.5 g NaCl, 1 g NH_4Cl. After autoclaving, add the following filter sterilized solutions: 1 ml of 1 M $MgSO_4 \cdot 7H_2O$, 0.5 ml of 2% thiamine-HCl in deionized H_2O, 10 ml of 20% glucose, deionized H_2O to 1 litre.
(d) **H medium**
  10 g bacto-tryptone, 5 g NaCl, deionized H_2O to 1 litre.

To prepare agar plates, add 15 g bacto-agar to 1 litre of medium before autoclaving, cool to ~60°C before pouring. To prepare top agar, add 0.7 g bacto-agar/100 ml and autoclave. Antibiotics should be added from stock solutions just before pouring or after liquid medium has cooled below 60°C.

ii Antibiotic solutions
(a) **Ampicillin:** The stock solution is 25 mg/ml of the sodium salt in water. Sterilize by filtration and store in aliquots at −20°C. The working concentration is 50 μg/ml.
(b) **Chloramphenicol:** The stock solution is 30 mg/ml in 100% ethanol. Store in aliquots at −20°C. The working concentration is 15 μg/ml.
(c) **Kanamycin:** The stock solution is 50 mg/ml in water. Sterilize by filtration and store in aliquots at −20°C. The working concentration is 70 μg/ml.

2.2 Introduction of phagemid into *E. coli* CJ236

2.2.1 Preparation of competent cells

Two methods are given for the preparation of cells competent to take up DNA. In the first method (*Protocol 1*) cells are prepared on the day they are to be used. When kept on ice these cells will retain their competence for about 24 h. In the second method (*Protocol 2*), cells are resuspended in 15% glycerol at the final step and frozen in liquid nitrogen or dry-ice/ethanol. If kept at $-70\,°C$ these cells will retain their competence for several months.

Protocol 1. Preparation of competent cells for use on same day

1. Streak MV1190 on a glucose-minimal medium agar plate. Grow at $37\,°C$ for one to two days until well-defined colonies appear. This plate can be stored in the refrigerator and used to inoculate liquid cultures for one or two months. The night before you plan to do the transformations and platings inoculate 20 ml of LB medium with a well-isolated colony. Grow overnight at $37\,°C$ with shaking.

2. Read the OD_{600} of the overnight culture. A 1/10 dilution gives an accurate reading.

3. To prepare competent cells, inoculate 40 ml of LB medium with enough of the overnight culture to give an initial absorbance reading of approximately 0.1. Incubate with shaking at $37\,°C$. The rest of the overnight culture can be saved on ice or in a refrigerator for use as a plating culture for titreing the helper phage.

4. When the culture reaches an OD_{600} of 0.8–0.9 (approx. 2 h; dilute 1 to 3 for accurate reading), harvest the culture by centrifugation at $0\,°C$. For easier resuspension, centrifuge only long enough to pellet the cells (e.g. 5 min at 5000 r.p.m. in a Sorvall Superspeed centrifuge). Carefully pour off the supernatant and drain well.

5. Using a pre-chilled pipette add 1 ml of cold 50 mM $CaCl_2$. Resuspend the cells **gently** by swirling. If it is difficult to obtain complete suspension they may be gently pipetted up and down with the chilled pipette. After a smooth suspension has been obtained, add an additional 19 ml of cold 50 mM $CaCl_2$ (final volume of 20 ml).

6. Hold the cells on ice 20–30 min.

7. Harvest the cells again by gentle centrifugation (step 4). Resuspend the cells **gently** in 1 ml 50 mM $CaCl_2$. Again, use a pre-chilled pipette to resuspend. Add an additional 3 ml of cold 50 mM $CaCl_2$ (final volume of 4 ml).

Protocol 2. Preparation of competent cells for storage at $-70\,°C$

1. Inoculate 200–250 ml of LB in a 500-ml Erlenmeyer flask to an OD_{600} of 0.1 from an overnight culture of MV1190, prepared as described in *Protocol 1*. Incubate with shaking at $37\,°C$.

2. When the OD_{600} reaches 0.9, harvest the culture by centrifugation at $0\,°C$.

Protocol 2. *continued*

For easier resuspension, centrifuge only long enough to pellet the cells (e.g. 5 min at 5000 r.p.m. in a Sorvall Superspeed centrifuge). Carefully pour off the supernatant and drain well. Resuspend the cells very **gently** in 50 ml of ice-cold 100 mM $MgCl_2$. If it is difficult to obtain complete suspension the cells may be gently pipetted up and down with a chilled pipette.

3. Again harvest the cells and drain the pellets well. Resuspend the cells **gently** in 10 ml of 100 mM $CaCl_2$ until a smooth suspension is obtained. Add 100 ml of 100 mM $CaCl_2$, mix, and keep the cells on ice 30–90 min.

4. Again harvest the cells and drain the pellets well. Resuspend **gently** in 12.5 ml of 85 mM $CaCl_2$ and 15% glycerol.

5. Immediately aliquot the cell suspension in 0.5–0.6 ml portions and freeze in either dry-ice/ethanol or liquid N_2. These cells will retain competence for at least 6–9 months if kept at $-70°C$. To use, simply thaw on ice and proceed according to *Protocol 3*.

The procedures described here will yield cells which can be transformed at an efficiency of 10^6–10^7 colonies/μg intact pTZ RF-I DNA which is sufficient for subcloning into the phagemids and isolating mutants from the *in vitro* mutagenesis reactions.

It is **very** important to keep the bacteria ice-cold during and after the first resuspension step in both protocols. All pipettes and solutions should be pre-chilled to ice-water (not just refrigerator) temperature. After chilling and addition of $CaCl_2$ the cells are very fragile. Care must be taken to ensure that resuspension and pipetting are **gentle**.

2.2.2 Transformation

Since the basis of the selection against the non-mutated strand in this procedure is the inviability of uracil-containing DNA in cells with an active uracil N-glycosylase, it is necessary to first prepare DNA containing uracil. This is done in the *dut* (dUTPase), *ung* (uracil N-glycosylase) strain CJ236. Because of the defective enzymes in this strain **all** DNA synthesized in it contains some uracil. First, the phagemid must be introduced into CJ236 by transformation (*Protocol 3*) or infection (Section 2.2.3). Once an isolate of CJ236 containing the phagemid is obtained, single-stranded uracil-containing phagemid DNA is generated by superinfecting the culture with M13K07 helper phage.

Protocol 3. Transformation of insert-carrying phagemid into CJ236

1. Grow CJ236 in the presence of chloramphenicol, and prepare competent cells (see *Protocols 1* and *2*).

2. Prepare double-stranded phagemid DNA (by standard plasmid miniprep methods) from an isolate containing the DNA to be mutated. DNA isolated from 1–5 ml culture is more than sufficient.

3. For each transformation, place 0.3 ml of competent cells in a cold 1.5 ml sterile polypropylene tube. Keep the tube on ice. If frozen competent cells are used, thaw them **on ice** and use 0.3 ml of cells per transformation. Do not allow the temperature of the cell suspension to rise above that of ice since competence will be rapidly lost.

4. To the competent cells, add 1–10 ng of DNA from a ligation reaction when cloning into pTZ, or 3–10 μl of an *in vitro* mutagenesis reaction after dilution with stop buffer as described in Section 2.5. Mix **gently**, and hold on ice for 30–90 min. DNA uptake appears to increase slowly from 30 to 90 min. No further increase is observed after that time.

5. Heat shock the cells by placing the tubes in a 42 °C water bath for 3 min, then return the cells to ice for at least 5 min.

6. The cells must now be grown for a period to allow expression of ampicillin resistance. To each transformation mix add 1 ml of LB and incubate with agitation for 1 h at 37 °C. Then spread 10–100 μl on to H plates containing ampicillin. Incubate the plates overnight at 37 °C. The colonies obtained from this transformation are used to prepare uracil-containing phagemid.

2.2.3 Infection

Infectious phagemid particles are produced when cells carrying the phagemid are superinfected with helper phage. The helper phage produces proteins which act on the f1 replication origin of the phagemid and cause single-stranded circular DNA to be synthesized and extruded from the cell encased in helper-phage coded proteins. These phage-like particles can infect cells carrying pili in exactly the same manner as the phage. Upon entry into the infected cell, a second DNA strand is made and the phagemid resumes its plasmid-like replication.

When wild-type M13 superinfects a host carrying a phagemid, neither are replicated to any extent. This phenomenon is called interference. M13K07 (11) has been constructed from an M13 mutant which partially bypasses this interference. Its own DNA replication has been partially disabled by insertion of DNA carrying the kanamycin resistance gene from Tn903 and the origin of replication of p15A. Together with the high-copy number of the phagemid DNA, this leads to packaging of the phagemid at the expense of the helper phage. The p15A origin of replication results in a concentration of phage products high enough to permit packaging of the phagemid DNA. The presence of kanamycin selects for cells infected with helper (11).

Protocol 4 outlines a procedure for infection of CJ236 by phagemid particles.

Protocol 4. Infection of CJ236 by insert-carrying phagemid particles

1. Grow a small culture (5 ml) of the transformed isolate in LB + ampicillin until the OD_{600} is between 0.1 and 0.4 (early log phase).

2. Make a 1/100 dilution of the M13K07 helper phage stock in TE (10 mM Tris, pH8, 1 mM Na_2 EDTA). Add $5 \mu l$ of this dilution to the culture and continue incubation at 37°C for 1 h.

3. Spin 1 ml of the culture in a microcentrifuge for 2–3 min to remove bacteria. Save the supernatant which contains the infectious phage particles.

4. Meanwhile, grow a 20 ml culture of CJ236 in LB + chloramphenicol at 37°C with aeration to an OD_{600} of 0.3–0.35. Inoculate this with $10 \mu l$ of the supernatant and continue incubation for 2 h.

5. Dilute the culture 1/10, 1/100, and 1/1000 in LB and spread $50 \mu l$ of each dilution on H or LB plates containing ampicillin. Incubate the plates at 37°C overnight. Do not incubate longer due to the growth of feeder (satellite) colonies.

6. From the plate with the least number of colonies, pick one and restreak it.

2.3 Isolation of uracil-containing phagemid particles

Protocol 5 describes the isolation of uracil-containing phagemids. It is essential that the uracil-containing template contains no extraneous DNA or RNA which may act as priming sites for second-strand synthesis. We have found two steps in the isolation of phagemid particles to be useful in minimizing endogenous primers—digestion with RNase A and an incubation in high salt buffer.

Protocol 5. Growth of uracil-containing phagemids

1. Streak out CJ236 containing pTZ with the desired insert on to an LB plate containing chloramphenicol. Grow at 37°C until distinct colonies appear.

2. Pick an isolated colony and place in 20 ml of LB containing $15 \mu g/ml$ chloramphenicol. Incubate with shaking at 37°C overnight.

3. Inoculate 50 ml of 2 × YT media containing ampicillin with 1 ml of the overnight culture of CJ236 containing phagemid. Incubate with shaking at

37°C. Only 30 ml of this culture will be used to isolate uracil-containing phagemid DNA. The remainder is surplus for reading ODs.

4. Grow to an OD_{600} of 0.3, which will take from 1 to 4 h. This corresponds to approximately 1×10^7 c.f.u./ml. Add helper phage M13K07 to obtain an m.o.i. of around 20, i.e. 20 phage/cell.

5. Incubate with shaking at 37°C for 1 h, then add $70 \mu l$ of the 50 mg/ml kanamycin stock. Continue incubation overnight at 37°C.

6. Transfer 30 ml of the culture to a 50-ml centrifuge tube. Centrifuge at $17000 g$ (e.g. 12 000 r.p.m. on the Sorvall SS-34 rotor) for 15 min. Transfer supernatant which contains the phagemid particles to a fresh centrifuge tube. Recentrifuge this at $17 000 g$ for 15 min at 0–4°C.

7. Transfer the second supernatant to a fresh polyallomer centrifuge tube and add $150 \mu g$ RNase A. See ref. 21 for details on preparing DNase-free RNase. Incubate at room temperature for 30 min.

8. To the phagemid-containing supernatant, add 1/4 vol. 3.5 M ammonium acetate/20% PEG-6000 and mix well. Incubate on ice for 30 min.

9. Collect the phagemids by centrifuging at $17 000 g$ for 15 min. Decant the supernant and discard. Drain the pellet well. Resuspend the pellet in $200 \mu l$ of high salt buffer (300 mM NaCl, 100 mM Tris-Cl, pH 8.0, 1 mM EDTA). Transfer the resuspended phagemids to a 1.5 ml microcentrifuge tube and chill on ice for 30 min. Centrifuge for 2 min to remove insoluble material and transfer supernatant to a fresh tube. Store at 4°C. DNA from this preparation should be extracted within one week.

In the event that a particular insert does not package efficiently, it may be possible to obtain phagemid particles by reversing the orientation of the insert. An example of this phenomenon has been described by McClary (12) with two clones each containing the same 2.3 kb HindIII fragment of bacteriophage λ. Phagemid DNA from the first clone is packaged whereas phagemid DNA from the second clone is not packaged and only helper phage DNA is present. Restriction enzyme analysis shows the clones to have the fragment in opposite orientation. The exact mechanism of non-packaging is unknown. It has also been reported that certain sequences will package either poorly or not at all in either orientation (11).

Since the phagemids obtained from CJ236 contain uracil in their DNA, they should survive far more readily in a bacterium without an active uracil N-glycosylase than in one with an active enzyme. If the phagemids were produced with a significant amount of uracil in their DNA, they should produce more ampicillin-resistant colonies in CJ236 than in MV1190 by a factor of approximately 10^4. A method for titreing the phagemid on these two strains is given in Protocol 6.

Protocol 6. Titreing phagemid

1. Grow overnight cultures of MV1190 and CJ236.

2. Add 0.5 ml of the MV1190 culture to 50 ml of $2 \times YT$. Add 1 ml of the CJ236 culture to another 50 ml of $2 \times YT$. Incubate with shaking at 37°C until the OD_{600} is 0.3 to 0.35. If one culture reaches this OD before the other, place it on ice.

3. Add 10 µl uracil-containing phagemid stock to each culture. Grow at 37°C for 2 h, with shaking.

4. Dilute the MV1190 culture 10-fold and 100-fold, and spread 50 µl of each dilution and the undiluted culture on to ampicillin-containing H or LB plates. Dilute the CJ236 culture 10^3-fold, 10^4-fold and 10^5-fold and spread 50 µl of these dilutions on to ampicillin-containing plates.

5. Incubate at 37°C overnight.

2.4 Extraction of phagemid DNA

After a suitable phagemid stock has been obtained the uracil-containing DNA must be purified as described in *Protocol 7*. The extractions are conveniently performed in standard 1.5 ml microcentrifuge tubes. Typically a few micrograms of DNA are obtained. Since only 0.2 µg is used in an *in vitro* mutagenesis reaction and this reaction should yield tens of thousands of transformants it is not necessary to isolate large amounts of DNA.

Protocol 7. Extraction of phagemid DNA

1. Extract the 200 µl phagemid stock (from *Protocol 5*, step 9) $2\times$ with an equal volume of neutralized phenol, $1\times$ with phenol/chloroform (1:1:1/48 phenol:chloroform:isoamyl alcohol), and several times with chloroform/isoamyl alcohol. Continue the chloroform/isoamyl alcohol extractions until there is no visible interface, then repeat once more. It is important to vortex each extraction for 1 min. The yield of DNA can be increased 30–50% by back-extracting each step: add 100 µl of TE (10 mM Tris, pH 8, 1 mM EDTA) to the first phenol extraction tube, vortex, add resultant aqueous phase to the next phenol extraction tube, vortex, and so on.

2. Pool the aqueous phases, add 1/10 vol. 7.8 M ammonium acetate and 2.5 vol. ethanol, mix then place at −70°C for at least 30 min.

3. Centrifuge 15 min in the cold, carefully remove the supernatant, wash the pellet gently with 90% ethanol, and resuspend the pellet in 20 µl TE. Avoid dissolving any residue that may cling to the side of the tube.

4. Transfer the dissolved DNA to a fresh tube. Run a small aliquot on an

agarose gel with a known amount of single-stranded DNA to determine the DNA concentration.

2.5 Synthesis of the mutagenic strand

After preparation of single-stranded uracil-containing template DNA from the phagemids, *in vitro* mutagenesis reactions are performed, and the reaction products transformed into the uracil N-glycosylase-containing strain MV1190. Synthesis from the single-stranded template DNA is primed with an oligonucleotide containing the sequence of the desired mutations(s). For discussions of various strategies for designing oligonucleotides for the purpose of inserting mutations, consult refs 13 and 14.

2.5.1 Oligonucleotide phosphorylation

We have found that the frequency of mutagenesis dropped threefold when the mutagenic primer was not phosphorylated. The oligonucleotide should be lyophylized after synthesis and resuspended in water at 10–20 pmol/μl. The procedure in *Protocol 8* has been taken from ref. 15 and been successfully used at Bio-Rad.

Protocol 8. Phosphorylation of the mutagenic primer

1. Prepare the following reaction in a sterile 500 μl microcentrifuge tube:

Component	Volume	Final conc. or mass
oligonucleotide	varies	200 pmol
1 M Tris, pH 8.0	3 μl	100 mM
0.2 M MgCl$_2$	1.5 μl	10 mM
0.1 M DTT	1.5 μl	5 mM
1 mM ATP (neutralized)	13 μl	0.4 mM
sterile water	varies	to total volume of 30 μl

2. Mix well.
3. Add 4.5 units of T4 polynucleotide kinase to the reaction mixture.
4. Mix and incubate at 37°C for 45 min.
5. Stop the reaction by heating at 65°C for 10 min.
6. Store frozen.

2.5.2 Template–primer annealing

Protocol 9 gives the procedure we use for annealing primer to template in the experiments described below. The uracil-containing DNA was obtained from pTZ18U that was altered by oligonucleotide mutagenesis to contain an amber mutation at codon 16 of the *lacZ'* fragment. The UGG codon (tryptophan) was converted to the chain terminating codon UAG. Since the phagemid contains an amber mutation in the *lacZ'* fragment and is therefore unable to complement *lacZ⁻* host strains, it yields white colonies on plates containing ampicillin, IPTG and X-gal (the amber mutation is not significantly suppressed by the *sup*E in MV1190). In some of the experiments described here the uracil-containing DNA was isolated from either M13mp11 or M13mp7 which contains the same amber mutation as described above.

A phosphorylated 16-base oligonucleotide was synthesized which will revert the mutation in the amber phagemid or amber M13 phage. This oligonucleotide has the sequence 5'pGGTTTTCCCAGTCACG3' and will revert the amber mutation UAG back to the wild-type UGG codon for tryptophan. Thus, blue colonies are produced on plates containing IPTG and X-gal. The underlined base is the one which is reverted in the amber phagemid or amber phage by this primer.

The molar ratio of primer to phagemid template in the reaction is between 20:1 and 30:1 for a 16-mer oligonucleotide and a 2860-base template, which are the sizes of the primer and template we use. We have found that higher ratios can interfere with subsequent ligation. Also, very high ratios of primer to template can result in a significant level of spurious priming from secondary hybridization of primer to the template. The annealing conditions described here have been optimized for use with our template and primer. It will be necessary to optimize conditions, which are expected to vary widely (2), for the particular oligonucleotide and template used.

Protocol 9 Annealing of primer to template

1. Prepare the following reaction mix in a $500\,\mu$l microcentrifuge tube:

Component	Volume	Final conc. or mass
uracil-containing DNA	varies	200 ng (0.3 pmol)
mutagenic oligonucleotide	varies	6–9 pmol
10 × annealing buffer[a]	$1\,\mu$l	20 mM Tris, 2 mM MgCl$_2$, 50 mM NaCl
water	varies	so that the total volume of the reaction is $10\,\mu$l

2. Prepare a second reaction mixture containing all the above ingredients except the primer.

3. Place the reaction mixtures in a 70°C water bath. Allow the reactions to cool in the water bath at a rate of approximately 1°C per min to 30°C over a 40 min period. After this, place the reactions in an ice-water bath.

[a] 10 × annealing buffer is 200 mM Tris, pH 7.5 (at 37°C), 20 mM MgCl$_2$, 500 mM NaCl.

Protocol 10 Second-strand synthesis

1. With the annealing reactions from *Protocol 9* still in the ice-water bath, add the following components in the order listed:

Component	Volume	Final conc. or mass
10 × synthesis buffer[a]	1 μl	0.4 mM each dNTP, 0.75 mM ATP, 17.5 mM Tris, 3.75 mM MgCl$_2$, 0.5 mM DTT
T4 DNA ligase	1 μl	3 units
T4 DNA polymerase	1 μl	1 unit

2. Incubate the reactions on ice for 5 min, then at 25 °C for 5 min, and finally at 37 °C for 90 min.

3. Add 90 μl of stop buffer (10 mM Tris (pH 8.0) 10 mM EDTA) to the reaction and stop the reaction by freezing. The reaction is stable at −20 °C for at least one month for use in the subsequent transformation.

[a] 10 × synthesis buffer is 5 mM each dATP, dCTP, dGTP and TTP; 10 mM ATP; 100 mM Tris, pH 7.9 (at 37 °C); 50 mM MgCl$_2$; 20 mM DTT.

2.5.3 Second-strand synthesis

The final step is synthesis of the second strand as described in *Protocol 10*. This involves two enzymes, T4 DNA polymerase and T4 DNA ligase, and occasionally T4 gene 32 protein (see Section 2.6). T4 DNA polymerase has intrinsic 5' to 3' polymerase and 3' to 5' exonuclease activities, but, unlike *E. coli* DNA polymerase I, will not hydrolyse DNA in the 5' to 3' direction (16). Significantly, it does not perform strand displacement (17, 18) and therefore will not readily remove the hybridized mutagenic primer. The Klenow fragment of *E. coli*, an enzyme commonly used for *in vitro* mutagenesis, does exhibit some strand displacement activity. We found a substantial difference in the efficiency of obtaining mutants using the two enzymes. For example, in one comparison experiment (*Figure 3*) using conditions for the Klenow polymerase which were typical of published procedures, we obtained 65% mutants in the T4 DNA polymerase reaction, but only 36% using the Klenow fragment. Incubation of the Klenow polymerase reaction for 18 h at 15 °C or increasing the proportion of ligase fourfold did not increase the mutation efficiency (4). The difference in the mutation efficiency may be due to the ability of the Klenow polymerase to peel off the mutagenic primer in a strand displacement reaction (19) and, hence, copy the template strand rather than retain the desired mutation.

T4 DNA ligase is used to ligate the newly synthesized DNA strand to the 5'-end of the oligonucleotide primer. The importance of this step is demonstrated in *Figure 4* which shows the results of mutagenesis reactions performed with various levels of ligase. The reaction products are elec-

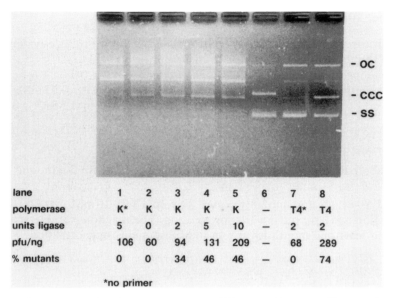

lane	1	2	3	4	5	6	7	8
polymerase	K*	K	K	K	K	−	T4*	T4
units ligase	5	0	2	5	10	−	2	2
pfu/ng	106	60	94	131	209	−	68	289
% mutants	0	0	34	46	46	−	0	74

*no primer

Figure 3. Comparison of T4 DNA polymerase and Klenow polymerase. The template used for this experiment was M13mp11 containing an amber mutation in the *lacZ'* gene. The primer reverts this mutation. Standard annealing and synthesis conditions were used for T4 DNA polymerase. The synthesis conditions for Klenow polymerase was a consensus from published procedures (22, 23, 24, 25, 26, 27). Twenty microlitre reactions containing 20 mM Tris–HCl, pH 7.5, 20 mM NaCl, 10 mM $MgCl_2$, 0.5 mM DTT, 0.5 mM each dNTP, 0.5 mM ATP, 0.2 μg of DNA, and 2 units of Klenow polymerase were performed. The gel is shown with the polymerase used, amount of ligase, p.f.u./ng of template and % mutation efficiency listed below each lane. Since it has been reported that the ratio of ligase to polymerase is important (23), we tested several different ratios of ligase to Klenow. Lanes 2–5 are the results with Klenow polymerase and varying amounts of ligase (0, 2, 5, and 10 units, respectively). Lanes 7 and 8 are the results with T4 DNA polymerase. Lanes 1 and 7 contained no primer. Lane 6 was a mixture of single-stranded template and M13 RF DNA. oc: open circles; ccc: covalently circular DNA; ss: single-stranded template; K: Klenow; T4: T4 polymerase. (Reprinted with permission of Eaton Publishing.)

trophoresed through a 1% agarose gel containing 0.5 μg/ml ethidium bromide. Under these conditions, it is possible to distinguish covalently closed circular DNA (ccc DNA) from incomplete circular DNA or nicked circular DNA (oc DNA); see Section 2.7 for a discussion of gel analysis of *in vitro* mutagenesis reaction products. The biological activity of the reaction products, measured by the number of plaques (in the case of M13 mutagenesis) or colonies (in the case of phagemid mutagenesis) obtained after transformation parallels the amount of ccc DNA produced in the reaction. As shown in *Figure 4*, in the absence of ligase we observed no ccc DNA, low biological activity, and a low frequency of mutation. Although in some experiments it is possible to get

42

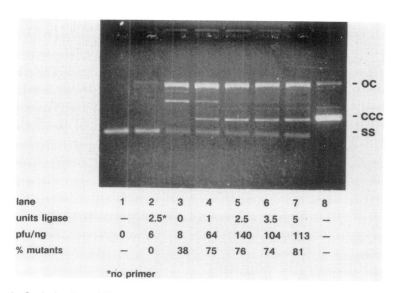

lane	1	2	3	4	5	6	7	8
units ligase	—	2.5*	0	1	2.5	3.5	5	—
pfu/ng	0	6	8	64	140	104	113	—
% mutants	—	0	38	75	76	74	81	—

*no primer

Figure 4. Optimization of ligase concentration. This experiment was performed with the same template and primer described in *Figure 3*. Standard annealing and conditions were used, except only 2–3 pmol of reverting primer was used and the amount of ligase was varied. The primer to template ratio was 20:1 for all reactions, except lane 2 which had no primer. The gel is shown with the amount of ligase, p.f.u./ng of template in the reaction, and % mutation efficiency listed below the corresponding lane. Lane 1 is a control in which the reaction mixture was incubated without the addition of T4 DNA polymerase and T4 DNA ligase. Lane 2 had no primer, but 2.5 units of ligase, added. Lanes 3–7 included 0, 1, 2.5, 3.5, and 5 units of ligase, respectively. Lane 8 is M13 RF DNA. oc: open circles; ccc: covalently closed circular DNA; ss: single-stranded template. (Reprinted with permission of Eaton Publishing.)

good mutagenesis without *in vitro* ligation (ref. 3, and T.A. Kunkel, pers. commun.), we recommend the inclusion of a ligation step.

The reaction conditions for DNA synthesis are (including the contributions from both annealing and synthesis buffers): 23 mM Tris (pH 7.4) 5 mM $MgCl_2$, 35 mM NaCl, 1.5 mM DTT, 0.4 mM dATP, dGTP, and dTTP, 0.75 mM ATP, plus the nucleic acids and enzymes. The reaction is initially started on ice for 5 min in order to stabilize the primer by initiation of DNA synthesis under conditions that favour binding of the primer to the template (3). The reaction then proceeds for 5 min at 25 °C and is finally continued for 90 min at 37 °C to permit complete synthesis of the second strand.

It is important to run a no primer reaction as a control to test for non-specific endogenous priming caused by contaminating nucleic acids in the template preparation. This test is important because endogenous priming may result in lowered mutation efficiency. In the absence of primer little single-stranded template is converted to RF-IV DNA, indicating that the template is not contaminated with impurities which can prime complementary strand DNA synthesis (*Figure 5*). At primer-to-template ratios of 20:1 to 30:1 we typically convert 50–80% of the template to primarily double-stranded material, and 10–50% of this material is converted to RF-IV DNA. Production of RF-IV in the absence of mutagenic primer is indicative of contaminated DNA template, and new template should be purified from a fresh phagemid stock. Failure to produce RF-IV DNA when primer has been added may be due to inactive T4 DNA polymerase, failure of the primer to hybridize to the template, or improperly set-up reactions.

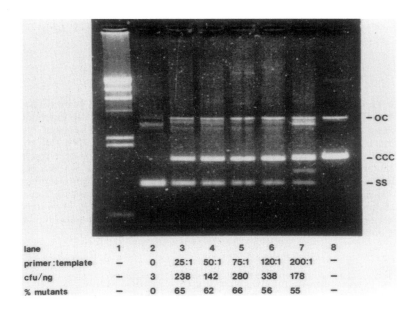

lane	1	2	3	4	5	6	7	8
primer:template	–	0	25:1	50:1	75:1	120:1	200:1	–
cfu/ng	–	3	238	142	280	338	178	–
% mutants	–	0	65	62	66	56	55	–

Figure 5. Effect of primer to template ratio. This experiment was performed with uracil-containing template obtained from pTZ constructed with an amber mutation in the *lacZ'* gene. The primer reverts this mutation. Standard annealing and synthesis conditions were used except that the amount of template was held constant at 200 ng and the amount of primer was altered to give varying ratios. The primer was the reverting primer and the template was the U-phagemid DNA. The gel is shown with the primer to template ratio, c.f.u./ng of template in the reaction, and % mutation efficiency listed below the corresponding lane. Lane 1 is *Hind*III-λ fragments as size standards. Lane 2 is a control in which no primer was added. Lanes 3–7 represent primer to template ratios of 25:1, 50:1, 75:1, 120:1, and 200:1, respectively. Lane 8 is pTZ plasmid DNA. oc: open circles; ccc: covalently closed circular DNA; ss: single-stranded template. (Reprinted with permission of Eaton Publishing.)

2.6 Use of T4 gene 32 protein

T4 gene 32 protein is a single-stranded DNA binding protein coded for by bacteriophage T4. One of its properties is to assist T4 DNA polymerase synthesis across regions of secondary structure in a single-stranded template (20). For most templates the polymerase should synthesize full-length copies in the absence of gene 32 protein. However, if upon gel analysis of the reaction product it appears that the enzyme is terminating synthesis prematurely, gene 32 protein may be added to the reaction to enhance synthesis. *Figure 6* shows the products of T4 DNA polymerase catalysed synthesis reaction using M13mp7 DNA as the template. This DNA contains a 24 bp inverted repeat in its polylinker region, therefore the single-stranded DNA forms a 'hairpin loop' at this point. Synthesis was primed a short distance upstream from this loop. We found that without substantial amounts of gene 32 protein, T4 DNA polymerase was unable to synthesize through this region. This experiment also demonstrates the effects of endogenous priming, since

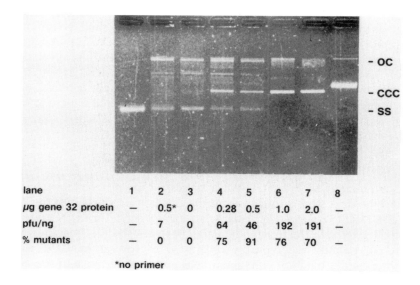

lane	1	2	3	4	5	6	7	8
µg gene 32 protein	—	0.5*	0	0.28	0.5	1.0	2.0	—
pfu/ng	—	7	0	64	46	192	191	—
% mutants	—	0	0	75	91	76	70	—

*no primer

Figure 6. Use of gene 32 protein to overcome secondary structure inhibition. This experiment was performed with template derived from M13mp7 phage which contained the same amber mutation as found in the U-phagemid DNA. Standard annealing and conditions were used, except only 2–3 pmol of reverting primer was used and T4 gene 32 protein was included. The gel is shown with the amount of gene 32 protein used, p.f.u./ng of template in the reaction, and % mutation efficiency listed below each corresponding lane. Lanes 3–7 included 0, 0.28, 0.5, 1.0, and 2.0 µg of gene 32 protein, respectively. Lane 2 included 0.5 µg of gene 32 protein, but no primer. Lane 1 is the single-stranded template. Lane is M13 RF DNA. oc: open circles; ccc: covalently closed circular DNA; ss: single-stranded template. (Reprinted with permission of Eaton Publishing.)

considerable DNA is copied into open circular form without the addition of primer (lane 2). This also explains the open circular DNA formed in the absence of gene 32 protein in lane 3.

To check if T4 gene 32 protein will assist second-strand synthesis, set up sufficient reactions to try 2 or 3 different concentrations of gene 32 protein. Just prior to adding the T4 DNA ligase and T4 DNA polymerase, add the gene 32 protein to the various reactions. We have found that approximately 1 μg of gene 32 protein is sufficient to stimulate synthesis on 200 ng of M13mp7 (7200 base-pairs). Therefore, 0.5, 1 and 2 μg gene 32 protein are reasonable amounts to try in the reactions. Proceed with the reaction and gel analysis of the reaction products. If the reaction is judged successful in the presence of the gene 32 protein, the reaction products may be transformed.

2.7 Gel analysis of the reaction products

The reaction products should be analysed on a 1% agarose gel in 1 × Tris-acetate buffer that contains 0.5 μg/ml ethidium bromide. The ethidium bromide binds to covalently closed circular (ccc) relaxed DNA (RF-IV DNA) and causes positive supercoils. This condensation causes the DNA to migrate more rapidly through the gel. Since the relaxed form of ccc DNA binds less ethidium bromide (which is positively charged) than the negatively super-coiled RF-I DNA, it often migrates slightly faster than RF-I DNA, but the two forms migrate very close together. The second-strand synthesis reaction results in the formation of relaxed ccc DNA; hence, a band migrating with or slightly ahead of RF-I plasmid DNA indicates successful conversion to the biologically active covalently closed circular RF-IV.

2.8 Transformation of the reaction mixture

Once the gel analysis indicates that a successful *in vitro* synthesis has been obtained the next step is to transform the reaction products into the *E. coli* strain MV1190 (*Protocol 11*). This strain has an active uracil N-glycosylase which will inactivate the uracil-containing parental strand, thus selecting for the mutant strand. We typically obtain 30 to several hundred colonies depending on the batch of competent cells and the efficiency of the *in vitro* synthesis reaction. Our competent cells usually yield around 500 colonies with 50 μl of the final suspension after transformation with 1 ng pTZ DNA. The number of colonies obtained in the no-primer reaction should be less than 20% of that obtained with primer. If results similar to these are obtained, colonies can be picked, purified and screened for those containing the desired mutant phagemid.

Protocol 11. Transformation of reaction products

1. Prepare competent MV1190 cells according to *Protocols 1* or *2*.

2. Transform $10\,\mu l$ of the reaction mixtures into $0.3\,\mu l$ competent cells following the procedure in *Protocol 3*. As controls, transform $3\,\mu l$ of the no primer reaction and 1 ng of either pTZ18U or 19U or the parental plasmid.

3. Spread 10 and $100\,\mu l$ of the expressed transformations on LB + ampicillin plates. The inclusion of IPTG and X-gal is not useful here since all the colonies will carry pTZ with an insert and will therefore all be white (or pale blue).

4. After allowing the bacterial suspension to be absorbed into the plates for about 10 min, invert them and incubate at 37°C overnight when colonies should be visible.

2.9 Analysis of transformants by DNA sequencing

Because of the strong selection against the uracil-containing non-mutant parental DNA strand of the *in vitro* mutagenesis reaction, typically more than 50% of the colonies obtained from the transformation of MV1190 carry mutant phagemids. This high efficiency of mutant production allows identification of mutants by direct DNA sequence analysis. This analysis is facilitated by an important property of phagemids, namely their existence in single-stranded form.

To prepare template for sequencing, grow a 50 ml culture from a purified colony of MV1190 containing a putative mutant as described in Section 2.3, except do not include chloramphenicol in the growth medium (chloramphenicol was used to ensure the presence of the F' plasmid in CJ236). Infect with M13K07 and concentrate phagemids as described Section 2.3. Extract the DNA as described in Section 2.4. This template may now be sequenced by standard Sanger dideoxy reactions. The pTZ phagemids contain the *lacZ'* region of pUC18 or pUC19 therefore the universal primer can be used if the mutation is within 250 bases of the 3'-end of the insert. Otherwise an oligonucleotide complementary to a position closer to the mutation may be used as sequencing primer.

References

1. Leatherbarrow, R. J. and Fersht, A. R. (1986). *Protein Engineering*, **1**, 7.
2. Kunkel, T. A. (1985). *Proceedings of the National Academy of Sciences of the USA*, **82**, 488.
3. Kunkel, T. A., Roberts, J. D., and Zakour, R. A. (1987). In *Methods in Enzymology* (ed. R. Wu), Vol. 154, p. 367.
4. Geisselsoder, J., Witney, F., and Yuckenberg, P. (1987). *BioTechniques*, **5**, 786.
5. McClary, J. A., Witney, R., and Geisselsoder, J. (1989). *BioTechniques*, **7**, 282.
6. Mead, D. A., Szczesna-Skorupa, E., and Kemper, B. (1986). *Protein Engineering*, **1**, 67.

7. Norander, J., Kempe, T., and Messing, J. (1983). *Gene*, **26**, 101.
8. Close, T. J., Christmann, J. L., and Rodriguez, R. L. (1983). *Gene*, **23**, 131.
9. Davis, R. W., Botstein, D., and Roth, J. R. (1980). *Advanced Bacterial Genetics*. Cold Spring Harbor Laboratory, Cold Spring Harbor, New York.
10. Miller, J. H. (1982). *Experiments in Molecular Genetics*. Cold Spring Harbor Laboratory, Cold Spring Harbor, New York.
11. Vieira, J. and Messing, J. (1987). In *Methods in Enzymology* (ed. R. Wu and L. Grossman), Vol. 153, p. 3.
12. McClary, J. (1989). *Molecular Biology Reports*, Bio-Rad Laboratories, No. 7, p. 5.
13. Craik, C. S. (1985). *BioTechniques*, **3**, 10.
14. Botstein, D. and Shortle, D. (1985). *Science*, **229**, 1193.
15. Zoller, J. J. and Smith, M. (1983). In *Methods in Enzymology* (ed. R. Wu, L. Grossman, and K. Moldave), Vol. 100, p. 468.
16. Huang, W. M. and Lehman, I. R. (1972). *Journal of Biological Chemistry*, **247**, 3139.
17. Massamune, Y. and Richardson, C. C. (1971). *Journal of Biological Chemistry*, **246**, 2692.
18. Nossal, N. G. (1974). *Journal of Biological Chemistry*, **249**, 5668.
19. Kornberg, A. (1980). *DNA Replication*, Freeman, San Francisco, California.
20. Huberman, J. A., Kornberg, A., and Alberts, B. N. (1971). *Journal of Molecular Biology*, **62**, 39.
21. Maniatis, T. Fritsch, E. F., and Sambrook, J. (1982). *Molecular Cloning: A Laboratory Manual*. Cold Spring Harbor Laboratory, Cold Spring Harbor, New York.
22. Adelman, J. P., Hayflick, J. S., Vasse, M., and Seeburg, P. H. (1983). *DNA*, **2**, 183.
23. Lewis, E. D., Chen, S., Kumar, A., Blanck, G., Pollack, R. E., and Manley, J. L. (1983). *Proceedings of the National Academy of Sciences of the USA*, **80**, 7065.
24. Nisbet, I. A. and Beilharz, M. W. (1985). *Gene Analytical Techniques*, **2**, 23.
25. Taylor, J. W., Ott, J., and Eckstein, R. (1985). *Nucleic Acids Research*, **13**, 8765.
26. Winter, G., Fersht, A. R., Wilkinson, A. J., Zoller, M., and Smith, M. (1982). *Nature*, **299**, 756.
27. Zoller, M. and Smith, M. (1982). *Nucleic Acids Research*, **10**, 6487.

3

Phosphorothioate-based site-directed mutagenesis for single-stranded vectors

JON R. SAYERS and FRITZ ECKSTEIN

1. Introduction

Oligonucleotide-directed mutagenesis allows the molecular biologist to alter a defined nucleotide sequence within a gene in a precise manner. This alteration, a mutation, may be a single mismatch, i.e. a transition or transversion, or it may involve the insertion or deletion of one or more bases. The effect of a mutation may manifest itself in the production of a new protein on expression of the modified gene. Most often the proposed active site of an enzyme is manipulated in this way. However, protein engineering is not the only application for site-directed mutagenesis (1, 2). The technique also has applications in the study of molecular recognition. Alterations in promoter sequences (3), repressor binding sites, and other regulatory sequences can be systematically introduced and the effect of the changes studied.

The basic approach to oligonucleotide-directed mutagenesis results in low mutational efficiency for the variety of reasons discussed in Chapter 1. As a consequence large numbers of putative mutant clones must be screened in order to obtain a clone with the required sequence. This involves a lot of effort and often the use of large amounts of radioactivity for hybridization screening or the sequencing of large numbers of clones. Thus, several strategies have been developed which aim to increase the efficiency of the basic site-directed mutagenesis procedure. Our method provides one such solution, while others are presented in Chapters 1 and 2 of this volume.

The phosphorothioate-based oligonucleotide-directed mutagenesis method overcomes the problems associated with transfection of a heteroduplex species (4–8). This is achieved by destroying the wild-type sequence opposite the mismatch primer and repairing the DNA *in vitro*. Thus, the resultant mutant DNA is transfected into competent cells as a fully complementary homoduplex species ensuring high mutational efficiencies which abolishes the need to plaque-purify primary transformants allowing the identification of mutants directly by sequencing a few transformants.

1.1 Development of the methodology

The phosphorothioate-based oligonucleotide-directed mutagenesis method is based on the observation that certain restriction endonucleases are incapable of hydrolysing phosphorothioate internucleotidic linkages (9, 10). Thus, double-stranded DNA containing phosphorothioate linkages in one strand only may be nicked in the non-substituted strand. In our mutagenesis procedure the mismatch oligonucleotide primer is annealed to the (+)strand of a single-stranded circular phage DNA. The primer is extended by a polymerization reaction in which one of the natural deoxynucleoside triphosphates is replaced by the corresponding deoxynucleoside 5'-O-(1-thiotriphosphate), dNTPαS (*Figure 1*). Thus, phosphorothioate groups are incorporated exclusively into the (−)strand of the newly synthesized RF-IV DNA. This results in a strand asymmetry which may be exploited.

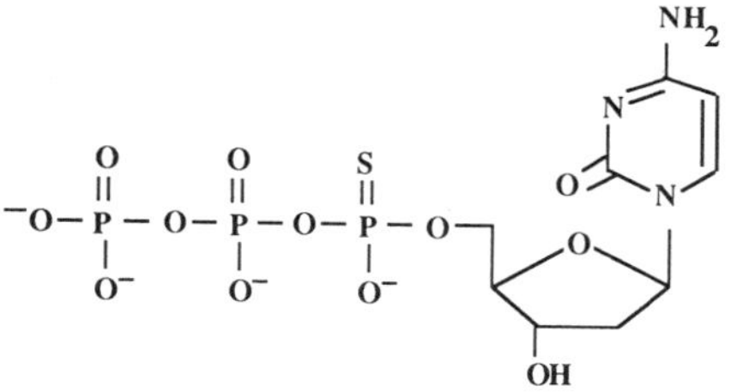

Figure 1. Structure of deoxycytidine 5'-O-(1-thiotriphosphate), dCTPαS. The sulphur atom replaces a non-bridging oxygen.

Reaction of such DNA with one of several restriction enzymes (for example, *Nci*I) produces a nick in the (+)strand as this enzyme is unable to cleave the (−)strand carrying a phosphorothioate group at the position of cleavage. The nick in the (+)strand is extended into a gap by reaction with a suitable exonuclease. Thus, the exonuclease digests away the wild-type sequence opposite the mismatch introduced by the primer. On repolymerization the gapped DNA is repaired using the mutant-carrying (−)strand as the template. The mutant sequence is now present in both strands of the DNA as a fully complementary homoduplex species. Transfection with this DNA produces mutational frequencies of the order of 85% (5–8). The polymerization, nicking, gapping, and repolymerization reactions (see *Figure 2*) must be performed carefully to obtain the highest possible mutational efficiencies.

2. Mutagenesis procedure

2.1 Preparation of single-stranded template DNA

The method shown in *Protocol 1* produces M13 single-stranded DNA suitable for use in the mutagenesis procedure without the need to perform DNase or

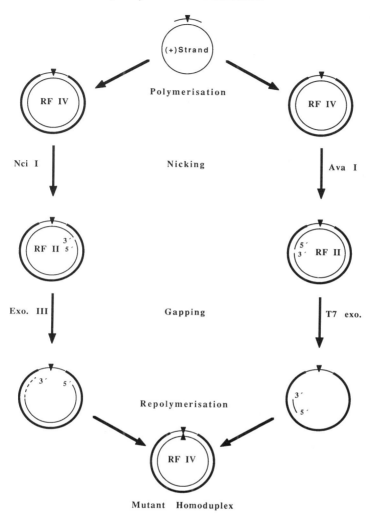

Figure 2. Schematic representation of the phosphorothioate-based mutagenesis method. Single-stranded DNA, annealed with a mismatch primer, is converted to RF-IV DNA using Klenow fragment, T4 DNA ligase, and dCTPαS. The newly synthesized (−)strand is shown by the heavy lines. The (+)strand of the phosphorothioate-containing DNA is then nicked by reaction with a restriction endonuclease such as *Nci*I or *Ava*I. The wild-type sequence is then digested away by either a 3′→5′ or a 5′→3′ exonuclease. A fully complementary homoduplex RF-IV molecule is generated on repolymerization.

RNase treatments or caesium chloride gradient purifications (6). The procedure includes two phage precipitation steps which are important for the preparation of suitable template DNA. This protocol provides at least 200–300 μg of single-stranded DNA template and may be conveniently scaled down by a factor of 3 as required.

Protocol 1. Preparation of template DNA

This protocol assumes that the phage carrying the insert of interest has already been prepared and its sequence determined.

1. Prepare an overnight culture of, e.g. SMH50 or TG1 cells, in 3 ml 2YT medium[a] by picking a colony grown on a glucose-minimal medium plate.

2. Prepare fresh cells by adding 1 drop of overnight culture from step 1 into 3 ml fresh 2YT and incubate at 37°C for 3 h in a shaker.

3. Pick a single plaque into 100 μl fresh cells and incubate overnight at 37°C.

4. At the same time set up another 3 ml overnight culture.

5. Inoculate 100 ml 2YT media (in a 250-ml Erlenmeyer flask) with 1 ml of cells from step 4 and grow with shaking at 37°C to OD_{600nm} 0.3.

6. Add the phage solution prepared in step 3. Continue incubation for 5 h.

7. Transfer the solution to centrifuge tubes and pellet the cells by centrifugation (e.g. 20 min at 23 000 g in a Sorvall centrifuge using a GSA rotor).

8. Immediately decant the supernatant and add 1/5 the volume of 20% PEG 6000 in 2.5 M NaCl. Allow the phage to precipitate for 30 min (or overnight) at 4°C.

9. Centrifuge at 3500 g for 20 min at 4°C. Discard the supernatant, remove traces of liquid with a screwed-up tissue or drawn-out pipette.

10. Add 1 ml TE buffer[b] (pH 7.4) and 9 ml H_2O to resuspend phage pellet.

11. Centrifuge at 3500 g for 20 min at 4°C. Transfer the phage containing supernatant to a clean centrifuge tube.

12. Add 2.2 ml 20% PEG in 2.5 M NaCl. Precipitate at 4°C for 30 min. Centrifuge at 3500 g for 20 min at 4°C.

13. Discard the supernatant, dissolve the phage pellet in 500 μl NTE buffer[c] and transfer to a sterile microcentrifuge tube.

14. Add 200 μl buffer equilibrated phenol, vortex for 30 sec spin briefly in a microcentrifuge. Transfer the aqueous (upper) layer to a fresh microcentrifuge tube.

15. Repeat step 14.

16. Add 500 μl water saturated diethyl ether vortex for 30 sec and spin briefly

in a microcentrifuge. Discard the upper (ether) layer. Repeat the process 3 more times.

17. Add $50\,\mu l$ 3 M sodium acetate (pH 6), mix and divide the solution into two microcentrifuge tubes.

18. Add $700\,\mu l$ absolute ethanol to each tube, cool to $-78\,°C$ in a dry-ice/propanol bath for 60 min. Centrifuge for 15 min in a microcentrifuge.

19. Discard the supernatant, add $700\,\mu l$ 70% ethanol, invert the tube to drain off the wash. Take great care not to dislodge the pellet.

20. Mark the tubes A and B, add $50\,\mu l$ DNA buffer^d to tube A. Resuspend the pellet by repeated vortexing. Transfer the buffer from tube A to tube B.

21. Add another $50\,\mu l$ of DNA buffer to tube A, vortex, spin briefly, and transfer contents to tube B. Tube B now contains $100\,\mu l$ buffer and DNA. Vortex to resuspend the pellet.

22. Take a $10\,\mu l$ sample, dilute to $1000\,\mu l$ and read the optical density on a UV spectrophotometer at 260 nm and 280 nm in a 1-ml quartz cuvette. The ratio of OD 260/280 should be 1.8 or higher, if not repeat the phenol extraction and following steps. 1 OD_{260} corresponds to approximately $37\,\mu g$ single-stranded DNA. Keep this sample as a standard for gel analysis, use $14\,\mu l$ diluted with stop mix.

^{a} 2YT contains 16 g tryptone, 10 g yeast extract and 5 g NaCl per litre, sterilized by autoclaving.
^{b} TE buffer contains 10 mM Tris–HCl, 1 mM EDTA.
^{c} NTE buffer contains 100 mM NaCl, 10 mM Tris–HCl, pH 8, 1mM EDTA.
^{d} DNA buffer contains 20 mM NaCl, 20 mM Tris–HCl, pH8, 1 mM EDTA.

It is important that the template is free from small pieces of RNA or DNA as such fragments may be capable of acting as primers in the polymerization reaction. A badly contaminated template may ultimately lead to the formation of wild-type double-stranded DNA and thus to reduced mutational efficiency. However, provided the infected cells are not allowed to grow for too long this problem is seldom encountered. We routinely perform a self-priming test on our single-stranded DNA (see *Protocol 2*). No significant amounts of polymerized material should be evident from this test.

Protocol 2. Self-priming test for single-stranded DNA

1. In a microcentrifuge tube mix
   $5\,\mu l$ 500 mM Tris–HCl, 1 mM EDTA (pH8)
   $5\,\mu l$ 500 mM NaCl
   $5\,\mu g$ single-stranded DNA template
   and adjust the final reaction volume to $23\,\mu l$ with H_2O.

Protocol 2. *continued*

2. Incubate at 70 °C for 5 min in a hot-water bath. Transfer immediately to a heating block at 37 °C and leave for 20 min. Place on ice.

3. Add

2.5 μl 4 × dNTP nucleotide mix.[a]
2 μl 500 mM Tris–HCl (pH 8)
5 μl 10 mM ATP
2 μl 100 mM MgCl$_2$
5 units each of DNA polymerase I and T4 DNA ligase.

4. Incubate at 16 °C overnight. Remove a 2-μl sample for gel analysis.

[a] The 4 × dNTP nucleotide mix contains dATP, dCTP, dGTP, and dTTP at 10 mM in each nucleotide, sterilized by filtration.

During all work with phage it is extremely important that all reusable equipment is sterilized. Phage-contaminated equipment should never be allowed to come into contact with any of the solutions used for transformation of the mutated DNA. Any contaminating wild-type phage will effectively render the procedure useless.

2.2 The mismatch oligonucleotide

The size of the mismatch primer obviously depends on the number of mismatches, or size of deletions or insertions which are required in the mutant sequence. The oligonucleotide used should have at least three bases to its 3'-side to protect it from the 3'→5' exonuclease proof-reading activity of the Klenow fragment (12). The number of nucleotides to the 5'-side of the mismatch depends on the quality of the Klenow enzyme used; it should be free of any residual 5'→3' exonuclease activity. For single- or double-base mismatches we routinely use oligomers of 18–22 nucleotides in length with the mismatch positioned toward the centre, such oligonucleotides bind quite specifically to their target sequence.

A deletion of 16 bases was achieved by using a 24-mer (12 bases either side of the deleted region). Similarly, an insertion of 16 bases was created by the use of a 40-mer as the mutagenic primer (12 bases either side of the inserted region). In both cases the mutational efficiency obtained was greater than 80% (7). It is possible to make quite large deletions (180 bp) using relatively short primers (T. Pope and M. Hetherington, pers. commun.). Such primers are readily available through automated DNA synthesis. The oligonucleotide sequence should be examined for obvious faults such as a high degree of self-complementarity as this could cause problems in the annealing step. If possible, the oligonucleotide should contain one or more dG or dC residues at its 5'-end as this serves as an anchor against strand displacement by virtue of the higher melting temperature of G:C base-pairs (11).

One final consideration is that the primer should not normally contain a recognition site for the restriction enzyme which is to be used in the nicking reaction. Such a site, if present, would lead to the linearization of the DNA during the nicking reaction, as the primer does not contain any phosphorothioate groups. However, it is possible to use a primer with such a recognition site provided that the primer is chemically synthesized so as to contain phosphorothioate groups at the positions required to protect it from the endonuclease. So far this approach has only been used with a primer containing an *Nci*I site, 5'-GGG*C*C*C-3'. The * indicates a phosphorothioate linkage introduced during solid phase synthesis. High efficiency mutagenesis has been achieved using such an oligonucleotide in our mutagenesis procedure (C. G. Hynes and M. Edge, pers. commun.).

2.3 Preparation of RF-IV DNA

The first step is the phosphorylation of the mismatch primer. The primer must be phosphorylated so that it can function as a substrate in the ligation reaction. This phosphorylation is catalysed by the enzyme polynucleotide kinase and requires ATP. Alternatively, the oligonucleotide may be chemically phosphorylated during solid phase synthesis, obviating the enzymatic kinase step. A portion of the phosphorylated primer is then combined with the template DNA. Although the phosphorylated primer may be stored for some time at −20°C we prefer to use freshly phosphorylated oligomers (see *Protocol 3*). The annealing step is carried out by heating approximately two molar equivalents of primer with the single-stranded template DNA in a high salt buffer. The denatured mixture is then placed immediately at the annealing temperature, for example 37°C, allowing the primer to anneal to the target sequence.

Protocol 3. Phosphorylation of mismatch oligonucleotide

1. Prepare the following mixture of reagents in a sterile 1.5-ml microcentrifuge tube. It is not necessary to use siliconized tubes:
   14 µl H$_2$O
   6 µl 500 mM Tris–HCl, 1 mM EDTA (pH8)
   2 µl Oligonucleotide primer (stock solution at 5 OD$_{260}$/ml)
   3 µl 10 mM ATP
   3 µl 100 mM MgCl$_2$
   2 µl 100 mM dithiothreitol (DTT)
   5 units of polynucleotide kinase (United States Biochemicals)

2. Vortex to mix the contents and centrifuge briefly to collect the solution.

3. Incubate at 37°C for 15 min in a heating block then at 70°C for 10 min in a hot water bath. Store on ice.

A high-quality Klenow fragment is required in the polymerization reaction (*Protocol 4*). It should be free of all detectable 5′→3′ exonuclease activity. Any such residual activity may serve to digest the mismatch primer from the 5′-end and could ultimately remove the mismatch. The polymerization reaction is usually carried out at 16°C in order to reduce strand displacement synthesis to a minimum. We usually obtain satisfactory yields of RF-IV DNA from such reactions after 16 h. Longer reaction times of up to 40 h do appear to increase the ratio of RF-IV to RF-II. Indeed, a polymerization started late on a Friday afternoon may safely be left until Monday morning when performed at 16°C. However, with some templates better yields of RF-IV DNA are obtained by performing the polymerization reaction at 37°C for 2–3 h, possibly due to the presence of secondary structures which are melted out at the higher temperature.

The ratio of Klenow fragment to T4 DNA ligase is an important factor. Too much Klenow favours strand displacement synthesis and is to be avoided. We have observed that a polymerase to ligase ratio of about 10 units Klenow to 15 units ligase produces the best results. Which of the deoxynucleoside phosphorothioates is used in the RF-IV preparation is dictated by the choice of restriction enzyme to be used in the nicking reaction. The restriction enzymes *Nci*I and *Ava*I have been used extensively and both require dCTPαS in the polymerization reaction in order to yield nicked DNA. Several other DNA polymerases such as those from phages T4 and T7, lacking a 5′→3′ exonuclease have become available recently. We have obtained good polymerizations by substituting the wild-type T7 polymerase from US Biochemicals (**not** their modified Sequenase™ version) for Klenow enzyme.

Protocol 4. Preparation of RF-IV DNA

Annealing of primer to template DNA

The annealing reaction is normally performed as follows for an oligomer of approx. 18–24 nucleotides.

1. Prepare the mixture in a sterile microcentrifuge tube by mixing
   10 μl 500 mM Tris–HCl, 1 mM EDTA (pH 8)
   10 μl 500 mM NaCl
   6 μl phosphorylated primer solution prepared in *Protocol 3*
   10 μg single-stranded DNA template (typically 2–5 μg/μl)

   and adjust the final reaction volume to 36 μl with H_2O.

2. Incubate at 70°C for 5 min in a hot-water bath.

3. Transfer immediately to a heating block at 37°C and leave for 20 min. Place on ice.

Polymerization

1. Add the following reagents to the template/primer mixture after the annealing mixture has been cooled on ice.

2.5 μl 10 mM dATP
2.5 μl 10 mM dGTP
2.5 μl 10 mM dTTP
2.5 μl 10 mM dCTPαS[a]
10 μl 10 mM ATP
10 μl 100 mM MgCl$_2$
3 μl 500 mM Tris–HCl, 1 mM EDTA (pH 8)
10 units of Klenow enzyme[b]
15 units of T4 DNA ligase

2. Adjust the volume of the reaction to ~80 μl with H$_2$O, mix and spin briefly.

3. Incubate for 16–40 h in a water-bath at 16 °C.

4. Heat inactivate at 70 °C for 10 min. Remove a 2-μl sample for gel analysis.

[a] The choice of phosphorothioate depends on which restriction enzyme is to be used in the nicking reaction. They are available from New England Nuclear.
[b] The Klenow enzyme supplied by New England Nuclear is adequate for this polymerization step. Alternatively Amersham International supply a kit comprising the enzymes and protocols required to perform the entire mutagenesis procedure.

2.4 Filtration through nitrocellulose

The correct performance of this step is extremely important to achieve the highest possible mutational efficiency. The RF-IV preparation usually contains small amounts of single-stranded DNA. This is, of course, wild-type DNA and is capable of producing wild-type plaques if transfected along with the mutated DNA. Any single-stranded DNA present at the end of the polymerization reaction is efficiently removed by a simple filtration through two nitrocellulose filters. Under the conditions described in *Protocol 5*, single-stranded DNA binds to the filter while polymerized material passes through. After filtration the DNA is subjected to a standard ethanol precipitation. It is important that the filtration step be carried out regardless of whether or not any trace of single-stranded DNA is apparent in the gel analysis.

Protocol 5. Nitrocellulose filtration and ethanol precipitation

1. Using forceps place first the rubber seal and then **two** nitrocellulose filters (**do not use autoclaved filters**) in the female part of the filter housing.[a]

2. Carefully wet the filter discs with 40 μl 500 mM NaCl and assemble the unit.

Protocol 5. *continued*

3. Attach a 2 ml disposable syringe to the outlet side of the filter unit using a short length of silicon tubing.

4. Add 6 μl 5 M NaCl to the polymerization reaction. Mix and apply to the inlet side.

5. Slowly draw the sample through the filter unit using the syringe plunger. You may need to tap the top of the housing gently in order to collect the filtrate. Not all of the solution volume is drawn through in the first stage.

6. Apply 50 μl 500 mM NaCl to the top of the filter unit and draw the wash through.

7. Carefully remove the filter unit and place the filtrate in a fresh sterile microcentrifuge tube. Transfer any remaining droplets with a micropipette.

8. Rinse the syringe with 50 μl 500 mM NaCl and combine with the filtrate.

9. Add 400 μl cold absolute ethanol, mix and chill at −78°C for 15 min.

10. Spin in a microcentrifuge for 15 min at ~14 000 r.p.m.

11. Discard the supernatant, a small pellet of salt/DNA should be visible.

12. Carefully add 400 μl of 70% ethanol (ethanol/water 7:3, v/v), invert the tube and check that the pellet is still stuck to the tube. Open the cap so that the liquid drains away. Be very careful not to disturb the pellet.

13. Carefully remove any traces of liquid with a drawn out pipette or a screwed-up tissue. Alternatively, dry on a SpeedVac concentrator for 2–3 min.

14. Resuspend the pellet in a total volume of 250 μl of nicking buffer (see *Protocol 6*).

[a] Filters were 13 mm in diameter, 0.45 μM pore size (SM11336) supplied by Sartorius. Filter units were from Millipore. The kit as supplied by Amersham International contains filter units which are operated by centrifugal force instead of a syringe system. This gives greater efficiency of DNA recovery.

2.5 Nicking reaction

Several restriction endonucleases may be successfully employed to produce a nick in the wild-type (+)strand although some appear to work better than others (5–8). *Nci*I and *Ava*I have been used most frequently in this laboratory. They both require that dCTPαS be used in the polymerization reaction. Such DNA then contains phosphorothioate linkages to the 5′-side of each dC residue in the (−)strand, except for the region covered by the mismatch primer. It is important that the nicking reaction is allowed to proceed to completion so that all the RF-IV DNA is converted to RF-II (nicked) DNA. Any RF-IV DNA left after the nicking step cannot be gapped by the exonuclease and lowers mutational efficiency. Nicking conditions for *Ava*I and *Nci*I are given in *Protocol 6*.

Protocol 6. Strand specific nicking of RF-IV DNA

A. RF-IV DNA containing dCMPS may be nicked by the enzyme *Nci*I[a] as follows:

1. Resuspend pellet from *Protocol 5* in 190 μl H_2O and add:

> 6 μl 500 mM Tris–HCl, 1 mM EDTA (pH 8)
> 25 μl 100 mM DTT
> 15 μl 100 mM $MgCl_2$
> 15 μl 500 mM NaCl
> *Nci*I (120 units)

2. Incubate at 37°C for 90 min.

3. Heat inactivate the enzyme at 70°C for 10 min. Keep a 6 μl sample for gel analysis.

B. RF-IV DNA containing dCMPS in the (−)strand may also be nicked with the enzyme *Ava*I.

1. Resuspend the DNA pellet (*Protocol 5* approx. 20 μg) in 160 μl H_2O and add:

> 13 μl 500 mM Tris–HCl, 1 mM EDTA (pH 8)
> 13 μl 100 mM DTT
> 25 μl 100 mM $MgCl_2$
> 37 μl 500 mM NaCl
> *Ava*I (70 units)

2. Incubate at 37°C for 180 min.

3. Heat inactivate at 70°C for 10 min. Keep a 6 μl sample for gel analysis.

4. The same protocol may also be followed for nicking with the enzymes *Ava*II[b] and *Ban*II[c].

---

[a] The double site in the polylinker region (CCCGGG 6247/8) of M13mp18 is not nicked at all by *Nci*I when dCMPS is present in the (−)strand. The nearest 'nickable' *Nci*I site to the polylinker is the one at 6838.
[b] *Ava*II may be used if dGTPαS was used in the polymerization reaction.
[c] Note that the *Ban*II recognition sequence GAGCTC present in the polylinker of M13mp18 is linearized by this enzyme. *Ban*II is thus unsuitable for a vector containing this site. However, the site GGGCTC in, e.g. M13mp2, mp7, mp8, mp9, or mp10, may be nicked with DNA containing dCMPS. Restriction enzymes were obtained Amersham International, Boehringer Mannheim, or New England Biolabs.

2.6 Gapping reaction

Exonuclease III was originally used in the method. It digests double-stranded DNA containing a free 3′-terminus in the 3′→5′ direction. Thus, a restriction site to the 3′-side of the target mutation relative to the (+)strand is required. Work with the M13mp18 vector has frequently used exonuclease III in conjunction with *Nci*I to degrade the (+)strand. *Nci*I has sites at positions 1924, 6247, 6248, and 6838. Interestingly, the double site in the polylinker region (CCCGGG 6247/8) is not nicked at all when dCMPS is present in the (−)strand (6). The nearest 'nickable' *Nci*I site to the polylinker is the one at 6838. This is important to note, as exonuclease III gaps nicked DNA at a rate of about 100 nucleotides per minute under the conditions shown in *Protocol 7*. A mismatch site in the middle of an insert of 1000 bases would therefore be approximately 1100 bases distant from the *Nci*I site at position 6838. In order to ensure that the mismatch (+)strand is completely digested a reaction time of about 20 min (allowing a safety margin) would be required. Exonuclease III may of course be used in conjunction with restriction endonucleases other than *Nci*I provided that the buffer conditions are adjusted accordingly.

Protocol 7. The gapping reaction using exonuclease III

The gapping reaction with exonuclease III is usually performed on *Nci*I nicked DNA. This protocol assumes that the nicked DNA is initially present in the *Nci*I nicking buffer.

1. Add $4\,\mu$l 5 M NaCl to the *Nci*I nicked DNA (*Protocol 6*).

2. Add ~300 units exonuclease III (New England Biolabs), vortex, spin briefly and incubate at 37°C for the time period required to gap through the mismatch (assume 100 bases per min with a safety margin of +50%).

3. Heat inactivate the exonuclease at 70°C for 15 min. Place on ice.

4. Remove an 8 μl sample for gel analysis.

The exonuclease III reaction is very sensitive to buffer conditions and the protocol should be followed exactly. The procedure described assumes that the nicked DNA is present in the *Nci*I reaction buffer. This is first converted to a higher salt concentration which is essential for reproducible gapping of the DNA with exonuclease III. We have found that exonuclease III gaps best in a buffer containing approx. 100 mM NaCl, 50 mM Tris–HCl (pH 8), 6 mM $MgCl_2$, and 10 mM DTT, using 15 units of exonuclease per microgram of nicked DNA. Gel analysis of the exonuclease III gapping reaction should reveal a distinct band whose electrophoretic mobility increases progressively with longer incubation time. Extended reaction is possible as exonuclease III can remove most of the (+)strand (7), resulting in a discrete band migrating very close to a marker of single-stranded DNA. However, we have observed that large amounts of DNA may be destroyed by contaminating endo-nucleases if the reaction is allowed to continue for too long.

The 5′→3′ exonucleases T7 and λ exonuclease produce gaps in nicked double-stranded DNA in the opposite direction to exonuclease III. Both are capable of removing almost all the nicked (+)strand under normal conditions, but unlike exonuclease III, discrete partially gapped DNA species cannot be detected by gel analysis. Commercial samples of these enzymes appear to be endonuclease free and prolonged incubation with either is usually possible. These exonucleases have been used in conjunction with restriction enzymes possessing a recognition site to the 5'-side of the insert. For most purposes the method given in *Protocol 8* produces a high degree of gapping; T7 exonuclease removes most of the (+)strand in less than 15 min with nicked M13mp18 as substrate. This latter exonuclease has emerged as the reagent of choice for the gapping reaction as it is available in concentrated form from US Biochemicals, is tolerant of variations in buffer conditions and does not destroy the (−)strand. The choice of exonuclease will depend on the position of the nick relative to the insert of interest. Whichever gapping system is chosen, it must be heat inactivated when complete.

The behaviour of the three exonucleases toward nicked, mismatch primed DNA is shown in *Figure 3*. A partial exonuclease III digest, allowed to continue for approximately 10 min, is shown together with T7 and λ exonuclease digests of the same DNA.

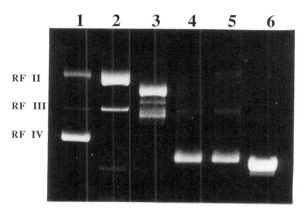

Figure 3. Gapping of nicked DNA with different exonucleases. Lane 1, RF-IV DNA prepared from a mismatch primer; lane 2, RF-II DNA prepared by reaction with *Aval* or *Ncil*; lane 3, partial gapping of RF-II DNA with exonuclease III, reaction terminated after 10 min; lanes 4 and 5, gapping of RF-II DNA with T7 and λ exonucleases, respectively; lane 6, marker of single-stranded circular M13 DNA.

Protocol 8. The gapping reaction using T7 or λ exonuclease

Both of these exonucleases appear to function in the buffers used for performing the nicking reaction.

T7 exonuclease

1. Add 10 units of T7 exonuclease (United States Biochemicals) per μg

Protocol 8. *continued*

double-stranded DNA present, i.e. 200 units for the *Ava*I nicked sample from *Protocol 6*.

2. Incubate at 37 °C for 30 min, heat inactivate at 70 °C for 10 min.

3. Place at 37 °C for 15 min then on ice. Remove a 14 μl sample for gel analysis.

λ **exonuclease** (New England Biolabs)

1. Add 4 units of λ exonuclease per μg double-stranded DNA present, i.e. 80 units for the *Ava*I nicked sample from *Protocol 6*.

2. Incubate at 37 °C for 120 min, heat inactivate at 70 °C for 10 min.

3. Place at 37 °C for 15 min then on ice. Remove a 14 μl sample for gel analysis.

2.7 Preparation of the mutant homoduplex: the repolymerization step

The gapped DNA must be repolymerized before transfection in order to obtain the best results. Transfection of gapped DNA seems to result both in low plaque yields and very low mutational efficiency. The repolymerization reaction shown in *Protocol 9* is carried out with DNA polymerase I (not Klenow fragment) in the presence of T4 DNA ligase. The polymerase should be of high quality and endonuclease free. The four natural deoxynucleoside triphosphates are used in this reaction. It may be carried out at 16 °C or 37 °C as is convenient. Note that the repolymerization reaction must contain ATP as well as dATP as was the case with the original RF-IV preparation.

Protocol 9. Repolymerization

1. To the gapped DNA solution prepared in *Protocols 7* or *8* add;
   20 μl 10 mM ATP
   5 μl 4 × dNTP mix, 10 mM in each nucleotide (*Protocol 2*).
   10 units DNA polymerase I
   10 units T4 DNA ligase

2. Incubate at 16 °C overnight or at 37 °C for 180 min.

3. Remove a 14 μl sample for gel analysis.

2.8 Transfection

One advantage of the phosphorothioate-based mutagenesis method is that specialized cell lines are not required for transfection or for growth of the template DNA. Any cell line suitable for the growth of M13 may be used. We have found such cell lines as SMH50 (13) and TG1 (14) are particularly useful in that they give consistently high transfection efficiencies; that is, the highest number of plaques per microgram DNA transfected. A transfection method is described in *Protocol 11* for competent cells prepared by the $CaCl_2$ method (*Protocol 10*). Obviously, other transfection protocols may be used if preferred (15, 16). Low plaque yield may be countered by precipitating the mutant DNA in 4 M NH_4OAc (17), or by passing the sample through Sephadex G-50 in a spun column (18). Both these methods remove nucleoside triphosphates which may be responsible for low plaque yield (19). Even if only little RF-IV can be detected by gel analysis after the repolymerization it is worth transfecting a sample of the DNA anyway.

Protocol 10. Preparation of competent cells

1. Add 3 ml of an overnight culture of, e.g. SMH50 or TG1 cells, to 100 ml of sterile 2YT media in a 250-ml Erlenmeyer flask.
2. Incubate in a shaker at 37 °C, allow to grow to an OD_{660nm} of 0.6.
3. Transfer cells to suitable sterile centrifuge tubes, cap and spin at ~3000 g for 15 min at 4 °C.
4. Discard the supernatant and resuspend the cells in a total volume of 50 ml pre-chilled 50 mM $CaCl_2$ solution (sterile).
5. Leave on ice for 30 min. Centrifuge as in step 3.
6. Discard the supernatant and resuspend the cells in a total volume of 20 ml pre-chilled 50 mM $CaCl_2$ solution (sterile). Store at 4 °C. Cells prepared in this manner may be used for up to one week but produce the best results ~24–48 h after this treatment.

Protocol 11. Transfection

1. Chill 5 sterile 10 ml polypropylene tubes on ice.
2. Add 300 μl of competent cells (*Protocol 10*) to each tube.
3. Dilute 20 μl repolymerized DNA solution to 50 μl with sterile H_2O.
4. Add 2 μl, 5 μl, 10 μl, and 20 μl aliquots of diluted DNA (step 3) to the competent cells.
5. To the 5th tube make a mock transfection with 20 μl sterile water used in diluting the DNA.
6. Swirl the tubes gently to mix the contents and place on ice for ~40 min.
7. Place 3 ml aliquots of molten top agar at 55 °C in a water bath (in sterile polypropylene tubes).
8. Combine 1400 μl fresh cells with 280 μl aqueous IPTG and 280 μl X-gal in dimethylformamide.[a]

Protocol 11. *continued*

9. To each aliquot of transformed competent cells (steps 4 and 5) add $270\,\mu$l of fresh cell mix from step 7.

10. Add 3 ml top agar to each mixture and plate out immediately on plates pre-warmed to 37 °C. Allow to set and invert.

11. Incubate overnight at 37 °C.

12. Pick 2–5 plaques for sequencing.

---

[a] Prepare the IPTG and X-gal solutions immediately before use. Dissolve 30 mg IPTG (isopropyl-β-D-thiogalactopyranoside) in 1 ml water, sterilize by filtration. X-gal (5-bromo-4-chloro-3-indolyl-β-galactoside) 20 mg, should be suspended in 1 ml dimethylformamide; do not attempt to sterilize!

---

3. Monitoring the procedure and debugging

Each stage of the phosphorothioate-based mutagenesis procedure is readily monitored by gel electrophoresis. This simple analytical technique is performed in 1% horizontal agarose slab gels (20). The gels and running buffer must be prepared so as to contain $0.4\,\mu$g/ml ethidium bromide, the running buffer should also contain 2-mercaptoethanol at $300\,\mu$l per litre, as this appears to guard against chemical nicking of the phosphorothioate DNA during electrophoresis which has occasionally been observed. We recommend that the novice checks each stage of the mutagenesis procedure before carrying out the next step. This saves time in the long run. Once the techniques have been mastered the analysis can be performed at the end of the sequence to save time. It is seldom worth transfecting a sample of hopefully mutant DNA if, for example, the nicking reaction failed! A gel eletrophoretic analysis is shown for the complete reaction sequence from polymerization to repolymerization in *Figure 4*. Note that it is advisable to run the samples along with a single-stranded DNA standard.

Gel analysis is very useful in fault diagnosis. A polymerization reaction which produces little or no RF-IV is readily detected. Such a result may indicate problems in the phosphorylation step or it may throw suspicion on the ligation reaction. Both of these reactions require ATP. A broad band in the vicinity of RF-II, with little or no RF-IV DNA present may also indicate that the polymerization reaction has not reached completion, a longer incubation period may help.

The use of a large excess of primer may result in the primer initiating polymerization at sites other than that desired, resulting in unligatable DNA as the Klenow fragment has no $5'\rightarrow3'$ exonuclease activity to digest away non-complementary ends (see *Figure 5*). Such non-specific priming is apparent by the polymerization reaction resulting in mostly RF-II-like material with bands of higher molecular weight DNA visible. The use of too little

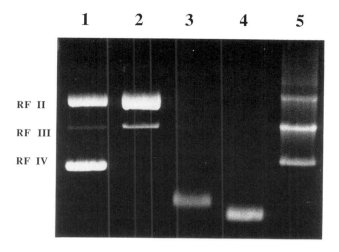

Figure 4. Agarose gel electrophoresis analysis of a complete mutagenesis procedure. Lane 1, polymerization reaction with dCTPαS; lane 2, *Ava*I nicking reaction; lane 3, gapping of RF-II DNA with T7 exonuclease; lane 4, marker of single-stranded circular M13 DNA; lane 5, repolymerization of gapped DNA.

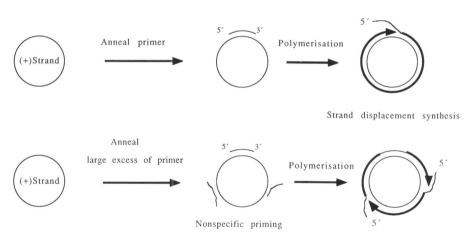

Figure 5. Shows two possible causes of unligatable RF-II-like species. In the upper diagram the correct amount of primer has been annealed but strand displacement results due to e.g. the use of an excess of Klenow enzyme over ligase. The growing (−)strand eventually displaces the 5'-end of the mismatch primer. In the lower diagram unligatable species arise due to annealing too much primer. Non-specific priming may occur, again leading to strand displacement synthesis.

primer is easily identified as large amounts of single-stranded DNA will be detected in the gel analysis if performed prior to the nitrocellulose filtration step.

The nicking reaction should convert all RF-IV to RF-II DNA, with the minimum of linearization. Should linearization result (RF-III DNA) then either the primer contains a recognition site for the restriction enzyme used or the wrong dNTPαS was employed in the polymerization reaction! It must be realized that not all the recognition sites of a restriction enzyme get cleaved at the same rate. The protocols given for *Ava*I, *Ava*II, *Ban*II and *Nci*I (*Protocol 6*) normally result in complete conversion of RF-IV to RF-II in the vectors we have used. Should the nicking reaction leave some RF-IV DNA it must be repeated or an extra aliquot of enzyme added to ensure complete conversion of RF-IV to RF-II DNA.

The gapping reaction is easily monitored in the case of the $5' \rightarrow 3'$ exonucleases as they usually gap the DNA almost completely, although faint traces of RF-II may remain. DNA gapped with either of these exonucleases will often run very close to a marker of single-stranded DNA and intermediately gapped species are not usually observed. The mobility of DNA that has been gapped with exonuclease III will depend on the duration of the gapping reaction. Gaps of less than about 400 bases may be very difficult to visualize. The repolymerization reaction usually yields reasonable amounts of RF-IV DNA although an increased amount of linearized material is observed. Should the only products be RF-II and RF-III DNA then the ligase or ATP has probably been left out of the reaction mixture. If only linear DNA is produced there may be a problem with oxidative cleavage of the phosphorothioate strand. The use of freshly prepared DTT in the nicking reaction usually solves this problem.

4. Modifications of the basic system

As with any general scheme there will be occasions when the procedure needs to be varied. The protocols given are based on primers in the 18–24 base size range. The use of primers designed to introduce deletions or insertions may require slightly different annealing conditions; a 40 mer designed to introduce an insertion of 16 nucleotides was subjected to an intermediate heating phase in the standard annealing protocol. Thus, after the 70 °C denaturation the mixture was placed at 55 °C for 20 min before being placed at 37 °C for a further 30 min. This resulted in production of mostly RF-IV DNA. Similarly, longer primers may require variations in the annealing step. The ratio of primer to template may also be varied over the range 1:1 to 6:1. However, at higher primer to template ratios the risk of non-specific priming increases, resulting in incomplete ligation (see *Figure 5*).

Another interesting feature of insertion or deletion mutagenesis is that the gapping reactions are slowed down or arrested by the single-stranded regions created by such a mismatch primer. In the case of a deletion the exonuclease must initially hydrolyse duplex DNA. Once the looped out deletion is reached the exonuclease has to gap a single-stranded region of DNA (the loop-out). The effect of an insertion primer is to place the loopout structure on the (−)strand, again the exonuclease has to pass through a shorter, single-stranded region. The exonucleases mentioned gap through such single-stranded regions quite poorly. However, DNA polymerase I which is used in the repolymerization reaction destroys the remaining wild-type strand allowing high mutational efficiency to be maintained. Gel analysis of such reactions may seem a little strange in that the fully gapped species usually observed with T7 or λ exonuclease may not appear. Instead a band with intermediate electrophoretic mobility will be seen.

One of the restriction endonucleases mentioned will usually prove useful for most practical purposes as we have a recognition site for *Nci*I to the 3′-side of the polylinker in M13mp18, for example. Whilst the restriction enzymes *Ava*I and *Ava*II have recognition sites to the 5′-side of the polylinker. *Ava*II requires dGTPαS in place of dGTP in the polymerization reaction, thus producing RF-IV DNA containing dGMPS in the (−)strand, otherwise the same procedure for nicking with this enzyme applies as for *Ava*I. *Ban*II may be used with several vectors such as M13mp2, mp8, or mp10. *Ban*II cannot be used with the vector M13mp18 as the GAGCTC site in the polylinker is linearized by this enzyme. However, these are not the only enzymes capable of giving nicked DNA. Other enzymes such as *Hin*dII, which requires dGTPαS, and *Pvu*I which requires dCTPαS during polymerization may also be used provided that the restriction enzyme is removed from the reaction by phenol extraction after completion of nicking (5). We have recently extended the range of restriction endonucleases suitable for use in the mutagenesis protocol by performing the reaction in the presence of ethidium bromide. In this way it was possible to use *Bgl*I, *Hpa*II, *Hin*dIII, and *Pvu*II in the mutagenesis procedure (8). We have also found that by incorporating two dNTPαS into the (−)strand of M13 a wider range of restriction enzymes may be used without the need for ethidium bromide (21). DNA prepared from a polymerization containing dATPαS and dGTPαS substituting for their natural counterparts produces DNA which is only nicked by the enzymes *Bam*HI and *Eco*RI. Similarly, DNA containing both dGMPS and dCMPS is nicked by the enzymes *Hha*I, *Hpa*II, *Msp*I, and *Pvu*II, whilst DNA containing dAMPS and dTMPS is nicked by *Hin*dIII. However, in most instances either *Nci*I or *Ava*I should prove effective in the procedure.

5. Scope and limitations of the procedure

The DNA to be mutated must be present in single-stranded form; for example, as an insert in one of the M13 derivatives. Alternatively, a vector containing a single-stranded phage origin of replication such as the pEMBL plasmids may be used (22). The recombinant DNA must have a restriction site for one of the restriction enzymes capable of producing nicked DNA. This site may be in the vector itself or the insert. The oligonucleotide used to introduce the mutation must not contain a site for the restriction enzyme, as it would simply linearize the DNA unless it was synthesized with phosphorothioate linkages at the appropriate positions.

Single and multiple base mismatches are equally accessible as are deletions and insertions. The size of a possible insertion is obviously limited by the length of reasonably pure oligonucleotide that may be prepared. Small insertions could also be used to enable easier cloning of an insert into an expression plasmid; for example, by creating different sticky ends compatible with the new vector. Frameshift mutations may be created simply by deleting or inserting 1 or 2 bases in the coding region of the insert.

Specialized host cells are not required either for amplification or transfection. Due to the high efficiency of mutagenesis it is usually sufficient to sequence a small number (3–5) of clones; hybridization experiments are not required.

References

1. Knowles, J. R. (1987). *Science*, **236**, 1252.
2. Shaw, V. W. (1987). *Biochemical Journal*, **246**, 1.
3. Min, T. T., Kim, M. H., and Lee, D-S. (1988). *Nucleic Acids Research*, **16**, 5075.
4. Sayers, J. R. and Eckstein, F. (1988). In *Genetic Engineering: Principles and Methods* (ed. J. K. Setlow), Vol. 10, pp. 109–22. Plenum Press, New York and London.
5. Taylor, J. W., Ott, J., and Eckstein, F. (1985). *Nucleic Acids Research*, **13**, 8765.
6. Nakamaye, K. L. and Eckstein, F. (1986). *Nucleic Acids Research*, **14**, 9679.
7. Sayers, J. R., Schmidt, W., and Eckstein, F. (1988). *Nucleic Acids Research*, **16**, 791.
8. Sayers, J. R., Wendler, A., Schmidt, W., and Eckstein, F. (1988). *Nucleic Acids Research*, **16**, 803.
9. Potter, B. V. L. and Eckstein, F. (1984). *Journal of Biological Chemistry*, **259**, 14243.
10. Taylor, J. W., Schmidt, W., Cosstick, R., Okruszek, A., and Eckstein, F. (1985). *Nucleic Acids Research*, **13**, 8749.
11. Fritz, H.-J. (1985). In *DNA Cloning—A Practical Approach* (ed. D. M. Glover), Vol. 1 pp. 151–63. IRL Press, Oxford.
12. Gillam, S. and Smith, M. (1979). *Gene*, **8**, 81.
13. LeClerc, J. E., Istock, N. L., Saran, B. R., and Allan, R. (1984). *Journal of Molecular Biology*, **180**, 217.
14. Carter, P., Bedouelle, H., and Winter, G. (1985). *Nucleic Acids Research*, **13**, 4431.
15. Hanahan, D. (1985). In *DNA Cloning—A Practical Approach* (ed. D. M. Glover) Vol. 1, pp. 109–35. IRL Press, Oxford.
16. Chung, C. T., Niemela, S. L., and Miller, R. H. (1989). *Proceedings of the National Academy of Sciences of the USA*, **86**, 2172.

17. Maniatis, T., Fritch, E. F., and Sambrook, K. (1982). In *Molecular Cloning: A Laboratory Manual*, pp. 461–2. Cold Spring Harbor Laboratory, Cold Spring Harbor, New York.
18. Maniatis, T., Fritch, E. F., and Sambrook, K. (1982) ibid., pp. 466–7.
19. Taketo, A. (1974). *Journal of Biological Chemistry*, **75,** 895.
20. Maniatis, T., Fritch, E. F., and Sambrook, K. (1982). *Molecular Cloning: A Laboratory Manual*, pp. 150–61. Cold Spring Harbor Laboratory, Cold Spring Harbor, New York.
21. Sayers, J. R., Olsen, D. B., and Eckstein, F. (1989). *Nucleic Acids Research*, **17,** 94–5.
22. Dente, L., Sollazzo, M., Baldari, C., Cesareni, G., and Cortese, R. (1985). In *DNA Cloning—A Practical Approach* (ed. D. M. Glove) Vol. 1, pp. 101–7. IRL Press, Oxford.

4

Site-directed mutagenesis using gapped-heteroduplex plasmid DNA

SUMIKO INOUYE and MASAYORI INOUYE

1. Introduction

Oligonucleotide-directed site-specific mutagenesis is the most effective and powerful method to create desired mutations at a specific site in a gene. The method involves the use of a short, synthetic oligonucleotide carrying the appropriate mutation, which acts as a primer of *in vitro* DNA synthesis on a complementary single-stranded circular DNA template. There are two basic methods to create a complementary single-stranded circular DNA template; one utilizes double-stranded plasmid (the plasmid method) and the other utilizes single-stranded M13 or phagemid vectors (the M13 method).

The M13 method originally developed by Zoller and Smith (1, 2) has been modified to obtain higher yields of mutants (see Chapters 1 to 3). Today, kits for the M13 method are commercially available with which one may get desired mutations with a yield of higher than 50%. However, one of the drawbacks of this method is that one usually has to reclone a target fragment into an M13 vector for mutagenesis. In addition, one has to move the mutated fragment back to the original expression plasmid after mutagenesis.

In contrast, the plasmid method allows one to directly use double-stranded plasmid DNA for mutagenesis. Although the yields of a desired mutation are usually lower (3–25%) with the plasmid method than with the M13 method, in practice the yields with the plasmid method are high enough to easily obtain desired mutation by a simple screening method. Therefore, the lower yields with the plasmid method are not considered to be a shortfall of the method. The detailed procedures of the plasmid method were previously described (3, 4, 5). In this chapter we describe improved procedures by utilizing a smaller amount of radioactivity to label a probe and by eliminating a purification using a Sephadex G-50 superfine column. This makes the plasmid method much simpler and easier.

2. The plasmid mutagenesis method

The plasmid method comprises four major steps described in detail in this section. *Figure 1* illustrates the general approach for the plasmid method. Only the desired region to be mutagenized is converted to the single-stranded form. As a result, substantial improvement in the mutant yield has been achieved in comparison to the earlier plasmid methods (3, 6, 7).

2.1 Preparation of plasmid DNA fragments I and II

A plasmid carrying a gene of interest has three restriction sites, A, B, and C (A and B can be for the same enzyme). It is desirable but not essential that the C site is in the drug-resistance gene of the plasmid, and the distance between the A and B sites is preferably less than 2 kb.

One-half of the plasmid DNA is digested at C, followed by treatment with the Klenow fragment of DNA polymerase I, or nuclease S1, or bacterial alkaline phosphatase (BAP). The Klenow or nuclease S1 treatment removes single-stranded sequences resulting from restriction endonuclease digestion at C. The resulting fragment I cannot confer drug resistance even if it forms a closed circular plasmid upon ligation, provided that C is in a drug-resistance gene. For example, the unique *Pst*I site in the ampicillin-resistance gene of pBR322 can be used as the C site. If C site digestion results in blunt cleavage, or is not in a drug-resistance gene, BAP treatment is essential to prevent self-ligation of fragment I. The C site should be at least several hundred base-pairs from both the A and B sites but need not be unique if the gap produced by C digestion is not too large and does not overlap with the gap produced by A plus B digestion.

In this chapter, as an example, the mutagenesis of a plasmid carrying the msDNA gene of *Myxococcus xanthus*, pMSSB (8) is described. *Figure 2* shows the restriction map of the plasmid. The C site used to create fragment I, is the unique *Xmn*I site in the β-lactamase gene as described in *Protocol 1*.

Protocol 1. Preparation of fragment I

1. Digest 5 μg pMSSB (7.7 kbp) DNA with *Xmn*I.

2. Extract the digest twice with phenol and three times with ether then ethanol precipitate the DNA.

3. Redissolve the DNA in 200 μl 10 mM Tris–HCl (pH 7.5), 0.1 mM EDTA.

4. Add 0.3 units bacterial alkaline phosphatase (Worthington) and incubate at 65 °C for 30 min.

5. Extract with phenol four times and with ether three times then ethanol precipitate.

6. Redissolve the DNA in 40 μl 0.1 × TE (TE is 10 mM Tris–HCl (pH 7.5), 1 mM EDTA).

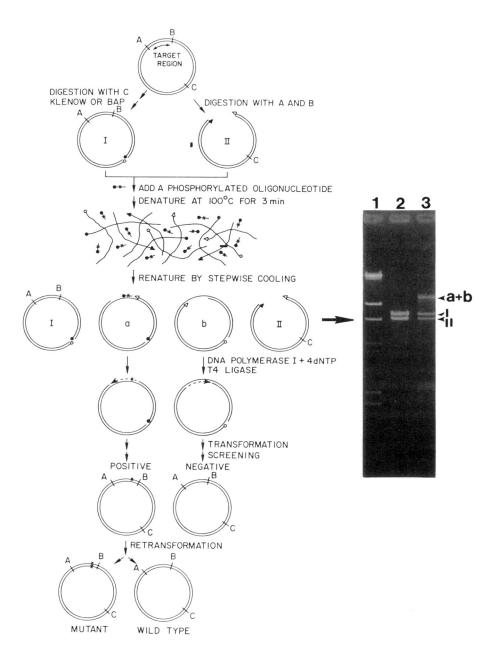

Figure 1. Schematic diagram depicting the steps involved in oligonucleotide-directed site-specific mutagenesis of a cloned circular double-stranded plasmid DNA and analysis of DNA after *Protocol 4*. For agarose gel electrophoresis fragments I and II were prepared as described in *Protocols 1* and *2*. Lane 1, a molecular weight standard (23, 9.4, 6.6, 4.4, 2.3, and 2.0 kilobases from the top to the bottom); lane 2, the mixture of fragments I and II (*Protocol 4*, step 2); lane 3, the same after *Protocol 4*, step 9. Modified from ref. 5.

73

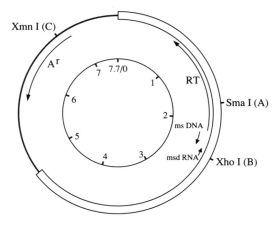

Figure 2. Restriction map of pMSSB used for the example.

Protocol 1. *continued*

7. Estimate the DNA concentration by analysing a 1 μl aliquot on a 0.7% agarose gel.

___

The other half of the plasmid DNA is digested at A and B to remove the target region. A small target region is desirable for higher mutant yield and for reducing the probability of heteroduplex formation with the mutagenic oligonucleotide at a site other than the target site. However, a reasonable mutant yield can be obtained even with a single-stranded region as large as 2 kb. The resulting fragment II is purified by 5% polyacrylamide gel electrophoresis. If A and B are the same enzyme, BAP treatment should be performed to prevent self-ligation of fragment II. If the target region is too large because of the lack of appropriate restriction sites, a smaller fragment may be isolated from the fragment AB and then added back in the denaturation mixture.

In *Protocol 2* the A and B sites used to create fragment II of pMSSB (*Figure 2*) are, *Sma*I and *Xho*I sites, respectively.

___

Protocol 2. Preparation of fragment II

1. Digest 5 μg of pMSSB with *Sma*I and *Xho*I.
2. Separate fragment II (7.0 kbp) from the 0.7 kbp fragment by 5% polyacrylamide gel electrophoresis and recover by electroelution (9).
3. Purify the DNA by extracting with phenol then with ether and precipitate with ethanol.
4. Redissolve the DNA in 40 μl of 0.1 × TE and estimate the DNA concentration as in step 7 of *Protocol 1*.

___

2.2 Design and preparation of oligonucleotides

For a base substitution mutation, an oligonucleotide of 15 bases in length (15-mer) is usually used in which the mismatched base is in the centre. To make a large deletion mutation, consisting of 10 to 50 bases, an oligonucleotide of approximately 24 to 30 bases (12 to 15 bases for each side of the deletion site) is used. An oligonucleotide containing the deleted sequence is then used as a probe for screening the mutants (negative screening).

Oligonucleotides can be purified by urea-polyacrylamide gel electrophoresis (9) or by using a column such as Oligonucleotide Purification Cartridges (OPC, Applied Biosystem). A purified oligonucleotide is dissolved in sterile 0.1 × TE [TE; 10 mM Tris–HCl (pH 7.5) and 1 mM EDTA] to a concentration of 20 pmol/μl. The oligonucleotide is then 5'-phosphorylated by treatment with T4 DNA kinase according to *Protocol 3A*.

Protocol 3. Phosphorylation of oligonucleotides

A. Preparation of mutagenic oligonucleotide primer

1. Mix the following components in a 1.5 ml microcentrifuge tube:

H$_2$O	Appropriate amount to make a final volume of 50 μl
10 × kinase buffer[a]	5 μl
Oligonucleotide (20 pmol/μl)	25 μl
1 mM ATP	1.5 μl
T4 polynucleotide kinase	5 U

2. Incubate the mixture at 37°C for 60 min.

3. Stop the reaction by incubating at 65°C for 10 min. Store at −20°C.

B. Preparation of oligonucleotide hybridization probe

1. Mix the following components:

H$_2$O	Appropriate amount to make a final volume of 30 μl
10 × kinase buffer[a]	3 μl
Oligonucleotide (20 pmol/μl)	1.5 μl
[γ^{32}P] ATP (3000 Ci/mmol)	30 μCi
T4 polynucleotide kinase	10 U

2. Incubate the mixture at 37°C for 60 min.

3. Chill on ice.

4. Apply 0.2 μl to PEI cellulose (4 × 10 cm; Brinkman) TLC in a 0.75 M potassium phosphate buffer (pH 3.7) system. The phosphorylated oligonucleotide hardly migrates from the origin whereas [γ-^{32}P] ATP and ^{32}Pi migrate much faster (the R_f of ATP is ~0.5 and the R_f of Pi is ~0.8).

Protocol 3. *continued*

Estimate the amount of radioactivity incorporated into the oligonucleotide. The incorporation of the radioactivity into the oligonucleotide should be more than 80% of the total activity used. However, even if the incorporation is less than 80%, the oligonucleotide can still be used as a probe.

ᵃ 10 × kinase buffer is 330 mM Tris–HCl (pH 7.5), 100 mM MgCl$_2$, 100 mM 2-mercaptoethanol.

2.3 Denaturation and renaturation

The procedures for formation and monitoring of heteroduplexes are given in *Protocol 4*. Equimolar amounts of fragments I and II are mixed with a 100-500-fold molar excess of synthetic oligonucleotide. The mixture is incubated at 100°C for 3 min to denature completely the DNA fragments. After the denaturation, the mixture is gradually cooled in a stepwise manner to allow the denatured DNA fragments to reanneal. During this reannealing procedure two new species of DNAs, DNA-a and -b, are formed in addition to the original fragments I and II. As shown in *Figure 1*, the mutagenic oligonucleotide hybridizes with only one of the two species of circular DNAs to form the heteroduplex since the oligonucleotide is complementary to only one of the two strands of the target gene.

The formation of the new circular DNAs should be confirmed by agarose gel electrophoresis (*Protocol 4*) before proceeding to the next step. As shown in *Figure 1*, after reannealing, a new band clearly appears at the position indicated by arrow a + b. This slower-migrating band consists of the new circular DNAs, DNA-a, and DNA-b, which are structurally almost identical. The position of the new band is dependent upon the size of the single-stranded region of the new circular DNAs. Sometimes, the new circular DNAs migrate at the same position as fragment I. In this case, the ratio of the intensity of band I to that of band II increases after reannealing. The migration of the new band becomes faster as the size of the single-stranded region increases. The open circular DNA also migrates faster when gel electrophoresis is carried out at a lower voltage for a longer time.

Protocol 4. Heteroduplex formation

1. Mix the following components in a 1.5-ml microcentrifuge tube:

0.1 × TE	To make a final volume of 30 μl
Fragment I	0.3 μg
Fragment II	0.27 μg
5′-phosphorylated primer	3.75 μl (step 3, *Protocol 3A*)

2. Remove 10 μl and hold on ice for agarose gel analysis alongside the sample from step 9 below.

3. Incubate the remaining mixture for 3 min in a boiling-water bath.

4. Take out tube, add $4 \mu l$ of $5 \times$ polymerase–ligase buffer[a] and incubate for another 1 min in a boiling-water bath.

5. Quickly transfer the tube to a 30°C incubator for 30 min.

6. Transfer the tube to a refrigerator (4°C) and incubate for 30 min.

7. Keep the mixture on ice for 10 min.

8. Spin the tube for 2 sec in a microcentrifuge at 4°C.

9. Remove $10 \mu l$ from the mixture and apply it to a 0.7% agarose gel with the sample from step 2.

10. Subject the samples to electrophoresis at 6 V/cm for 2–3 h in TAE buffer.[b]

11. Stain the gel with ethidium bromide.

[a] $5 \times$ polymerase–ligase buffer: 500 mM NaCl, 33 mM Tris–HCl (pH 7.5), 40 mM MgCl$_2$, 5 mM 2-mercaptoethanol.
[b] TAE buffer: (40 mM Tris-acetate, pH 8.0, 20 mM sodium acetate, 2 mM EDTA).

The position of the new band comprising the heteroduplexes ((a + b); see *Figure 1*) is variable depending on the electrophoretic conditions. It is important to run electrophoresis at a high temperature for the separation of the open circular DNA from fragments I and II.

The existence of the new band is crucial for the next step. If the new band cannot be detected one should **not** proceed to the next step. In this case, the following points should be checked:

(a) Was the ratio of fragment I to fragment II approximately = 1?

(b) Was the mixture maintained at 100°C throughout denaturation? The denaturation can be monitored by gel electrophoresis just following step 4 above. The denatured DNAs should migrate much faster than the double-stranded species.

(c) If DNA fragments have a high GC content or the heteroduplex band is not clearly formed for some other reasons, the concentration of the polymerase–ligase buffer may be increased or decreased.

2.4 Primer extension and transformation

After confirming the formation of the heteroduplex DNAs, these are incubated with the Klenow fragment of DNA polymerase I, T4 ligase, and four deoxyribonucleotide triphosphates as described in *Protocol 5*, part A. This treatment converts the open circular DNAs to closed circular DNAs. When fragment I is prepared by BAP treatment in *Protocol 1*, they still remain open circular. In the case of DNA-a, the oligonucleotide also serves as a primer for the polymerase. After the reaction, the mixture is used for transformation (*Protocol 5*, part B).

When A and B are the same restriction sites the addition of ligase can be eliminated to avoid self-ligation of fragment II. The elimination of ligase reduces the number of transformants by a factor of 2–3 and reduces the mutant yield if A and B are different, but may increase the mutant yield if A and B are the same.

Protocol 5. Primer extension, ligation, and transformation

A. **Extension/ligation step**

1. Mix the following components in a 1.5 ml microcentrifuge tube:

H$_2$O	Appropriate amount to make a final volume of 20 μl
The remaining reaction mixture from step 8 of *Protocol 4*	10 μl
2.5 mM each of dTTP, dCTP, dATP, and dGTP	4 μl
10 mM ATP	2 μl
Klenow enzyme	5 U
T4DNA ligase	1 U

2. Incubate the mixture at 14°C overnight.

B. **Transformation step**

3. Add 1 μl of the reaction mixture to 20 μl of CaCl$_2$-treated competent cells prepared according to ref. 9.

4. Keep the mixture on ice for 30 min followed by incubation in a 42°C water bath for 2 min.

5. Add 0.4 ml of L broth (9) to the mixture and incubate the final mixture at 37°C for another 60 min.

6. Plate 0.1 ml on each of four agar plates containing 50 μg/ml ampicillin (or an appropriate antibiotic).

7. Incubate the plates at 37°C overnight.

The number of transformants per plate should be 200–400.

2.5 Screening and confirmation of the mutations

Transformants carrying the mutated plasmid are screened by colony hybridization using the same synthetic oligonucleotide which was used as the mutagen. The oligonucleotide labelled with ^{32}P according to *Protocol 3B* is used as the probe for the screening. It forms an unstable heteroduplex with the wild-type plasmid DNA, but forms a perfectly complementary stable

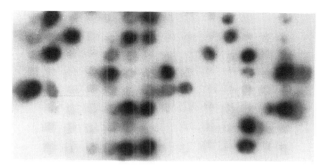

Figure 3. Colony-hybridization analysis using ^{32}P-labelled oligonucleotide as the probe. Colonies were grown on the Whatman 3MM filter paper and DNAs were denatured, fixed on the filter paper and hybridized with ^{32}P-labelled probe. Dark spots indicate positive colonies.

homoduplex with the mutant plasmid DNA. Therefore, the hybridization temperature is chosen for each oligonucleotide so that stable duplexes are formed only with the mutant but not with the wild-type plasmid DNA.

Three methods for screening of mutants were described previously (5):

(a) *Pick-and-screen method*;

(b) *Filter assay method I* using a circular Whatman No. 1 Chromatography paper;

(c) *Filter assay method II* using a circular nitrocellulose filter.

Although there are some advantages for each method, the *pick-and-screen method* is probably the most dependable way to detect mutations and is described in *Protocol 6*. Since the yields of mutations are usually 3–25%, it is not particularly labour-intensive to pick only 100–200 transformants and to screen them by hybridization in order to isolate at least a few mutants. A square plate, 100×100 mm, is used to pick colonies. All the agar plates used in this section should contain an appropriate antibiotic according to the plasmid vector used.

The filter paper is prepared as follows: Whatman 3MM filter paper is cut to 8×8 cm squares on which lines are pencilled every 5 mm to make grids containing a total of 196 small squares for spotting cells. The paper is sterilized by autoclaving, and carefully placed on a pre-warmed 100×100-mm square plate. Care should be taken not to trap air bubbles under the filter paper.

Protocol 6. Colony hybridization of transformants

A. Filters

1. Transfer the individual transformants on to the sterile filter on an agar plate, as described above, and a master plate using sterile applicator sticks

Protocol 6. *continued*

(the ends of which should be smoothed by filing). Spot cells evenly on the filter by gently turning the applicator sticks on the filter.

2. Incubate the plates face-up for the filter plate and face-down for the master plate at 37°C overnight.

3. Remove the filter papers from the plates in order to fix the DNA on to the filter paper.

4. Float the filter on 10 ml of 0.5 N NaOH in a Petri dish (diameter 15 cm) at room temperature for 7 min. Remove the solution by aspiration and add 10 ml of 0.5 N of NaOH gently from the bottom to float the filter paper again. Incubate for another 7 min.

5. Exchange the solution in the Petri dish with 20 ml of 0.5 M Tris–HCl (pH 7.4) and submerge the filter paper. Incubate with gently swirling for 3.5 min at room temperature. Repeat.

6. Exchange the solution with 30 ml of 1 × SSC (150 mM NaCl, 5 mM sodium citrate, pH 7.2), and gently swirl the Petri dish for 3.5 min at room temperature. Repeat.

7. Exchange the solution with 50 ml of 95% ethanol, and vigorously swirl the Petri dish for 7 min at room temperature. Repeat.

8. Dry the filter on Whatman 3MM paper for 20–30 min at room temperature.

9. Sandwich the filter paper between Whatman 3MM paper, wrap it in aluminium foil, and bake it for 2 h at 80°C *in vacuo*.

B. Hybridization

10. Put filters in a heat-sealed bag or zip-lock bag.

11. Prepare 4 ml hybridization solution per 8 × 8-cm Whatman 3MM square filter by mixing the following components:

50 × Denhardt's solution[a]	0.4 ml
30 × NET[b]	0.8 ml
20% Dextran sulphate (Pharmacia)	2 ml
10% sodium dodecyl sulphate	0.2 ml
[32]P-labelled oligonucleotide[c] plus H$_2$O to make up 4 ml	

12. Hybridize for 16–18 h at an appropriate temperature calculated for each oligonucleotide.[d]

13. Wash the filters with 100–200 ml of 6 × SSC three times for 20 min at r.t.

14. Wash the filters in 95% ethanol twice for 10 min. Air-dry the filters and place on cardboard. Also fix radioactive position markers on the cardboard.

15. Using an intensifying screen, expose Kodak XAR-5 or similar X-ray film to the filters for 2–5 h.

16. After developing the film, positive colonies should appear as dark spots (*Figure 3*). Orient the filters on the developed film using the radioactive positive markers. Transfer the filter markings to the film.

17. Pick positive colonies individually and isolate plasmid DNAs from 1 ml cultures as described by Birnboim and Doly (10).

18. Retransform competent cells under dilute DNA conditions with the plasmids and pick 20–30 colonies for each positive isolate.

19. Rescreen the positive colonies using the same method as described above.

[a] 50 × Denhardt's solution (11): Ficoll 1 g, Poly(vinylpyrrolidone) 1 g, BSA (Sigma, Fraction V) 1 g, H_2O to 100 ml.
[b] 30 × NET: 4.5 M NaCl, 0.45 M Tris–HCl (pH 7.5), 0.03 M EDTA.
[c] ^{32}P-labelled oligonucleotide prepared according to *Protocol 3B* is added at a final concentration of 5×10^5 c.p.m./ml.
[d] The hybridization temperatures are calculated by the following equation empirically derived for synthetic oligonucleotide hybridization (12): [number of A's plus T's × 2 + number of G's plus C's × 4] − 4. If the calculated temperature is higher than 55 °C, use 50 °C or add formamide for hybridization. Example: For a 15-mer oligonucleotide consisting of 7 (A + T) and 8 (G + C), the hybridization temperature is 42 °C = (7 × 2 + 8 × 4) − 4.

If washing the filters at room temperature does not give clear positives because of high background (in particular, for those which are done at high hybridization temperatures), filters can be washed at 5 °C lower than the hybridization temperature.

Since each positive transformant may contain the mutant plasmid as well as the wild-type plasmid it is important to isolate the mutant plasmid by retransformation. From one or two retransformed clones, plasmid DNA is purified and sequenced to confirm the mutation caused by the mutagenic oligonucleotide. The DNA sequencing can be performed directly on the double-stranded DNA using an appropriate oligonucleotide primer without transferring the mutated region to M13 vectors. DNA sequencing using a supercoiled DNA as a template is well-established and widely used (13). Commercially available DNA sequence kits using a modified T7 DNA polymerase can be used for DNA sequencing of plasmids.

If a restriction site is lost or created as a result of the mutation, restriction site analysis of the plasmid DNAs from the transformants can also be used for the primary screening for the mutant.

3. Concluding remarks

Oligonucleotide-directed site-specific mutagenesis is one of the most powerful techniques in recombinant DNA technology, and has now become a routine method in molecular biology laboratories for the study of the molecular mechanism of gene regulation, gene structure, protein structure and function and engineering of new enzymes. The plasmid method using double-stranded DNA described in this chapter has been dramatically improved and is now comparable to the M13 method in terms of the yield of mutations and the ease of the manipulations.

References

1. Zoller, M. J. and Smith, M. (1982). *Nucleic Acids Research*, **10**, 6487.
2. Zoller, M. J. and Smith, M. (1983). In *Methods in Enzymology* (ed. R. Wu, L. Grossman, and K. Moldave), Vol. 100, pp. 468–500. Academic Press, New York.
3. Vlasuk, G. P. and Inouye, S. (1983). In *Experimental Manipulation of Gene Expression* (ed. M. Inouye), pp. 291–303. Academic Press, New York.
4. Morinaga, Y., Franceschini, T., Inouye, S., and Inouye, M. (1984). *Bio/Technology*, **2**, 636.
5. Inouye, S. and Inouye, M. (1987). In *Synthesis and Applications of DNA and RNA* (ed. S. A. Narang), pp. 181–206. Academic Press, New York.
6. Wallace, R. B., Schold, M., Johnson, M. J., Dembek, P., and Itakura, K. (1981). *Nucleic Acids Research*, **9**, 3647.
7. Inouye, S., Soberon, X., Franceschini, T., Nakamura, K., Itakura, K., and Inouye, M. (1982). *Proceedings of the National Academy of Sciences of the USA*, **79**, 3438.
8. Hsu, M.-Y., Inouye, S., and Inouye, M. (1989). *Journal of Biological Chemistry*, **264**, 6214.
9. Sambrook, J., Fritsch, E. F., and Maniatis, T. (ed.) (1989). *Molecular Cloning: A Laboratory Manual* (2nd ed). Cold Spring Harbor Laboratory, Cold Spring Harbor, New York.
10. Birnboim, H. C. and Doly, J. (1979). *Nucleic Acid Research*, **7**, 1513.
11. Denhardt, D. T. (1966). *Biochemical and Biophysical Research Communications*, **23**, 641.
12. Suggs, S., Wallace, R. B., Hirose, T., Kawashima, E., and Itakura, K. (1981). *Proceedings of the National Academy of Sciences of the USA*, **78**, 6613.
13. Toneguzzo, F., Glynn, S., Levi, E., Mjolsness, S., and Hayday, A. (1988). *Biotechniques*, **6**, 460.

5

Phosphorothioate based double-stranded plasmid mutagenesis

DAVID B. OLSEN and FRITZ ECKSTEIN

1. Introduction

The ability to specifically modify a nucleotide sequence has facilitated many detailed studies of structure function relationships of DNA and proteins (1–3). To this end oligonucleotide-directed mutagenesis using single-stranded DNA vectors has become almost routine (see refs 4–7 and Chapters 1 to 3 in this volume). The high mutational yields using these well-established methods allow for a rather facile screening of the altered genotype.

One of the limitations of M13-based procedures is that the gene of interest must be subcloned into one of the M13 vectors described by Messing (8). Alternatively, subcloning may be avoided if the gene is present in a phagemid vector which is a plasmid which contains an origin of replication for single-stranded phage [(9); see Chapters 1 and 2]. However, working with these constructs sometimes requires special experience in order to obtain single-stranded DNA suitable for mutagenesis.

We have recently extended the phosphorothioate-based mutagenesis methodology described in Chapter 3 to double-stranded DNA (10). The basis of this procedure is the production of a specific region of single-stranded DNA to which an oligonucleotide containing one or multiple mismatches can anneal. Following repolymerization, all wild-type DNA is enzymatically hydrolysed, whereas the mutant homoduplex remains intact. After transformation of competent cells mutational efficiencies can be obtained which reach those described for protocols using single-stranded DNA templates.

2. Overview of the methodology

The first step in carrying out highly efficient plasmid mutagenesis is the site-specific nicking of the double-stranded substrate in the vicinity of the desired mutation. We have taken advantage of the observation that many restriction endonucleases are unable to hydrolyse both strands of DNA when incubated in the presence of ethidium bromide. This reaction is not strand-specific and a mixture of products are obtained which contain the nick in one of the two strands (*Figure 1, B*). This nick serves as a starting point for digestion by a non-processive exonuclease possessing either a 3′→5′ or a 5′→3′ hydrolytic activity. Two different gapped products are obtained depending on which strand contained the nick. For the example given in *Figure 1* we have chosen exonuclease III.

The product of the gapping reaction which leaves a small stretch of single-stranded DNA complementary to the mutant oligonucleotide is considered 'productively gapped' (*Figure 1, C*). After a strand-selective hybridization of the mutant primer, the gaps in both strands are filled by polymerization using the Klenow fragment and three dNTPs along with one dNTPαS. The choice of dNTPαS is dependent upon the restriction enzyme to be used in the next reaction (*Figure 1, D*). As seen in *Figure 1* the desired mutation is present as a heteroduplex in only one of the various plasmid DNA species present after polymerization.

One of the difficulties in carrying out highly efficient plasmid mutagenesis is the specific destruction of all the wild-type sequence. The introduction of phosphorothioate groups into the mutant strand during the polymerization step allows for strand-selectivity. Subsequent reaction with a restriction endonuclease which is unable to cleave phosphorothioate containing DNA (7, 11–13) will hydrolyze all the wild-type DNA strands. This includes the linearization of roughly 50% of the DNA population which is present, due to 'unproductive gapping' (*Figure 1, E*). We have chosen the enzyme *Pst*I for the case cited here.

The heteroduplex DNA containing phosphorothioates in the mutant strand at the recognition site of the restriction enzyme is only nicked. At this point the situation is very similar to the mutagenesis procedure described in Chapter 3. The nick created by the restriction endonuclease can be taken as a starting point for exonucleolytic digestion (*Figure 1, F*). The mutant strand is then used as template for repolymerization to form the homoduplex carrying the desired changes in both strands (*Figure 1, G*). The DNA is then directly transformed using appropriate cell lines. Experiments which consisted of several single- and double-base substitutions into the polylinker or *lacZ*α gene of the plasmid pUC19 have yielded efficiencies up to 70–80% (10).

2.1 Preparation of double-stranded DNA templates

The success of this method is highly dependent upon the quality of the DNA used as starting material. The amplification procedure outlined in *Protocol 1* will yield up to 2 mg of plasmid DNA. It is important that the DNA be free of contamination by small pieces of RNA which could compete with the mutant oligonucleotide for binding to the gapped template. Alternatively, these pieces of RNA could bind to the mutant oligonucleotide directly thus decreasing the amount of primer available for hybridization. It is recommended that after

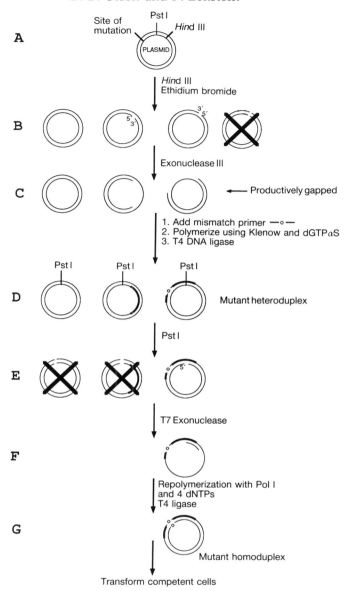

Figure 1. Schematic representation of the oligonucleotide-directed plasmid mutagenesis technique. A, pUC19 double-stranded covalently closed circular DNA; B, products from *Hind*III reaction; C, products from exonuclease III reaction; D, mutant heteroduplex; E, products from *Pst*I nicking/linearization reaction; F, products from T7 exonuclease reaction; G, repolymerized mutant homoduplex. The symbol 'O' represents the mutation within the mismatch oligonucleotide. Heavy lines indicate the area where phosphorothioates have been incorporated. The plasmid that has been linearized is crossed off because it is unable to transform.

CsCl-ethidium bromide purification, $2\,\mu$g of the DNA be analysed by agarose gel electrophoresis. If large molecular weight RNA is still present a second RNase digestion should be carried out. This should be followed by phenol-extraction and spin-dialysis using a Centricon-30 microconcentrator (Amicon) as outlined in *Protocol 2*. Even if RNA is not visible when a large amount of DNA is applied to the gel, we still recommend that $\sim\!100\,\mu$g of the plasmid be spin-dialysed and an aliquot of the DNA removed for the mutagenesis procedure.

Protocol 1. Preparation of plasmid DNA from amplified cultures[a]

1. Prepare overnight culture by using a sterile toothpick to stab a single colony and place into 3 ml LB media[b] containing $50\,\mu$g/ml ampicillin.

2. Inoculate 3×2 litre flasks each containing 500 ml LB/Amp medium with 1 ml of overnight culture.

3. Shake at 37°C until an absorbance of 0.5 at 600 nm.

4. Add 1 ml of chloramphenicol solution (300 mg/3 ml ethanol) to each flask.

5. Incubate for 10–16 h at 37°C, with vigorous shaking.

6. Centrifuge the cells at 3000 g for 10 min at 4°C.

7. Resuspend the pellet in 30 ml cold suspension-buffer (10% sucrose, 6 mM Tris–HCl, pH 8) and bring to 40 ml with suspension buffer.

8. Divide the suspension equally into four Sorvall SS34 or equivalent centrifuge tubes.

9. Add $400\,\mu$l of lysozyme solution (50 mg lysozyme/5 ml suspension buffer) to each tube, mix gently and incubate on ice for 5 min.

10. Add 1.5 ml 250 mM EDTA solution (pH 8 with NaOH) and $50\,\mu$l RNase A solution[c], mix gently and place on ice for 5 min.

11. Add 10 ml H_2O to each tube and mix gently.

12. Add 10 ml of the SDS solution[d] and mix gently.

13. Place tubes on ice for 5 min then centrifuge at $\sim\!48\,000\,g$ for 120 min at 4°C.

14. Carefully remove the supernatant from each tube and precipitate the DNA by incubation on ice for 180 min with one-third volume of 30% polyethylene glycol 6000 solution in 1.5 M NaCl.

15. Centrifuge for 20 min at $\sim\!48\,000\,g$ (4°C) and carefully remove the supernatant. The inside of the tube is dried using a drawn out pipette.

16. Dissolve the DNA in 5 ml of H_2O with the help of a pipette, and bring the volume to 11.5 ml with H_2O.

17. Add 11.85 g CsCl and $100\,\mu$l ethidium bromide (10 mg/ml) and centrifuge at 170 000 g for 20 h at 10°C.

18. After the centrifuge has stopped (no brake) the lower band visible in the tube is harvested using a syringe.

19. Dilute the DNA solution with 2 vol. of TE buffer (10 mM Tris–HCl (pH 8), 1 mM EDTA) and extract with CsCl saturated isoamyl alcohol until all traces of ethidium are removed from the lower phase.

20. Precipitate the DNA by the addition of 3 vol. of absolute ethanol and incubate on ice for 30 min.

21. Centrifuge the DNA for 10 min at $\sim$12 000 g and discard the supernatant.

22. Wash the pellet with 70% ethanol and dry the pellet using a vacuum.

23. Resuspend the DNA in 0.5–1 ml distilled H_2O and determine the concentration by diluting 10 μl to 1 ml. Place this solution into a 1-ml quartz cuvette and record the A_{260}. Multiply this value by 5 to give the concentration of the stock DNA in μg/μl. Keep this sample, as it can serve as a marker when the samples are visualized by agarose gel electrophoresis.

[a] See ref. 22.
[b] LB media: 10 g bacto-tryptone, 5 g bacto-yeast, 10 g NaCl per litre adjusted to pH 7.5 with NaOH.
[c] RNase A solution; 10 mg/ml H_2O and placed in boiling-water bath for 2 min.
[d] SDS solution 120 mM Tris–HCl (pH 8), 30 mM EDTA, 2% (v/v) Triton X-100.

Protocol 2. Extraction and spin-dialysis procedure

1. Add 200 μl buffer-equilibrated-phenol (see ref. 23) and vortex for a minimum of 30 sec.

2. Briefly spin the tube and remove the aqueous layer with a pipette.

3. Repeat steps 1 and 2 using 200 μl water-saturated $CHCl_3$/isoamyl alcohol (24:1).

4. Repeat steps 1 and 2 with 1 ml water-saturated diethylether.

5. Remove final traces of ether by heating the tube with the cap open at 70°C for 10 min.

6. Dilute the sample with 2 ml of distilled water.

7. Add the sample to a Centricon-30 filtration apparatus (Amicon) and spin for 20 min at, for example, 3500 r.p.m., using a Sorvall SS34 fixed-angle rotor.

8. Repeat steps 6 and 7 two more times.

9. Collect the sample, and transfer solvent containing DNA (50–60 μl) to a sterile 1.5-ml microcentrifuge tube.

2.2 Preparation of mutant heteroduplex

2.2.1 Site-specific nicking of the double-stranded DNA

The enzymes *Hin*dIII and *Eco*RI are present at either the 3'- or 5'-end of a number of multiple cloning sites found in a variety of plasmid vectors. In *Protocol 3* we describe optimized conditions for the nicking of plasmid DNA using these two enzymes. In addition, a large number of other restriction endonucleases have been reported to nick double-stranded DNA in the presence of ethidium bromide (14–18). So this step is not limited to these two enzymes, but their use should prove the most universal.

Protocol 3. Nicking of plasmid

1. Add the following reagents to a 1.5 ml sterile microcentrifuge tube:

   ~20 μg plasmid DNA
   100 μl 2 × *Hin*dIII buffer[a]
   10 μl *Hin*dIII 20 units/μl
   20 μl ethidium bromide 500 μg/ml

2. Bring the volume of the solution to 240 μl using sterile distilled water, briefly vortex and spin.

3. Incubate at 30°C for 60 min.

4. Remove 2 μl for agarose gel electrophoresis.

5. Purify DNA according to *Protocol 2*

Alternatively, the DNA can be nicked using *Eco*RI with the conditions given below.

1. Add the following reagents to a 1.5-ml sterile microcentrifuge tube:

   ~20 μg plasmid DNA
   120 μl 2 × *Eco*RI buffer[b]
   30 μl *Eco*RI (20 units/μl)
   72 μl ethidium bromide 500 μg/ml

2. Bring the volume of the solution to 260 ml using sterile distilled water and briefly vortex the tube to mix contents.

3. Incubate at 16°C for 15 h.

4. Remove 2 μl for agarose gel electrophoresis.

5. Purify DNA according to *Protocol 2*.

[a] *Hin*dIII buffer contains 100 mM Tris–HCl (pH 7.4), 20 mM MgCl$_2$, 100 mM NaCl.
[b] *Eco*RI buffer contains 200 mM Tris–HCl (pH 7.4), 200 mM NaCl, 800 μM CoCl$_2$ (see ref. 23).

Ideally, the enzyme chosen for this initial nicking step should be present only once in the vector or insert and it cannot overlap the region where the mutant oligonucleotide is to anneal. In addition, it is advantageous to have the nicking site within several hundred base-pairs from the location of the mutant oligonucleotide. However, it is unimportant if this site is upstream or downstream from the mutation. One of the reasons for wanting the nicking site relatively close to the mutational site is that the time for exonuclease digestion can be shortened. It is also advantageous to create as small a gap as possible so it is less likely that the mutant oligo will anneal to another single-stranded DNA region.

2.2.2 Production of a specific region of single-stranded DNA

There are several considerations which govern the choice of exonuclease used to produce the region of single-stranded DNA. The enzyme must create a gapped area inclusive of the sites of the desired mutation and the second restriction endonuclease. As mentioned previously, the enzyme should hydrolyse the DNA non-processively and in a defined direction. A non-processive enzyme is required so that the time of incubation determines the size of the gap. In our hands the best results were obtained using exonuclease III (see *Protocol 4*). However, T5 exonuclease can also be used for this step (J. Sayers and F. Eckstein, manuscript in preparation).

Protocol 4. Gapping with exonuclease III

1. Adjust the volume of the nicked plasmid solution to 80 μl with water and add:

   10 μl of freshly prepared 10 × E9 buffer[a]
   4 μl 1 M NaCl
   1 μl Exonuclease III 100 units/ml (New England Biolabs)

2. Briefly vortex the tube to mix the contents and spin to collect the solution in the bottom.

3. Incubate at 37°C for the time required for digestion past the mismatch.[b]

4. Remove 2 μl for agarose gel electrophoresis.

[a] 10 × E9 buffer, 100 mM Tris–HCl (pH 8), 100 mM MgCl$_2$, 600 mM NaCl, 70 mM DTT.
[b] See *Protocol 7* from Chapter 3 for information regarding the time of the gapping reaction using exonuclease III.

After the gapping reaction the sample can be analysed by agarose gel electrophoresis before proceeding further. The gapped DNA should run on the gel somewhere between the nicked and linear plasmid DNA bands used as markers (see Section 3). It is very important to note that the enzyme should not be allowed to gap to such an extent that the 'unproductively gapped' species (*Figure 1*) incorporate phosphorothioate groups at the site of the second restriction endonuclease. A band which migrates much faster than the linear plasmid DNA marker would be indicative of over-digestion. If this does occur the wild-type sequence would be protected from being linearized, thus causing a dramatic decrease in the mutational efficiency. It is equally important that a sufficient degree of gapping occurs so that the mutant oligonucleotide can anneal to the single-stranded DNA.

2.2.3 The mutant oligonucleotide

The types of mutations that one can introduce by this method such as point mutations, insertions and deletions have been outlined in Chapters 1 to 4, which also contain general information about designing a mutant oligonucleotide. In addition, the oligonucleotide must be synthesized such that it is complementary to the single-stranded region remaining after digestion. The sequence of the oligonucleotide will vary depending whether a $3' \rightarrow 5'$ or a $5' \rightarrow 3'$ exonuclease is chosen for the gapping reaction. One example of the versatility with respect to the choice of exonucleases and primers is given in *Figure 2*.

Prior to the annealing the mutant oligonucleotide must first be phosphorylated according to *Protocol 5*. Approximately two molar equivalents of phosphorylated primer are then heated with the template DNA as described in *Protocol 6* and slowly cooled to room temperature to allow hybridization to occur.

Protocol 5. Phosphorylation of the mismatch oligonucleotide

1. Add the following reagents to a sterile 1.5 ml microcentrifuge tube.

   6 μl 100 mM MgCl$_2$
   6 μl 500 mM Tris–HCl (pH 8)
   6 μl 100 mM β-mercaptoethanol
   6 μl ATP 10 mM
   60 μl mismatch oligonucleotide   (5 A$_{260}$ Units/ml)
   10 units polynucleotide kinase

2. Adjust the volume to 100 μl using sterilized distilled water, briefly vortex the tube to mix the contents and spin to collect the solution in the bottom.

3. Incubate at 37°C for 30 min.

4. Heat inactivate the enzyme at 70°C for 15 min.

Protocol 6. Annealing of the mutant primer to the gapped plasmid[a]

1. Add the following reagents to the solution containing the gapped DNA from *Protocol 4*:

   3 μl phosphorylated oligo (3 A$_{260}$ units/ml, *Protocol 5*)
   10 μl 1 M NaCl

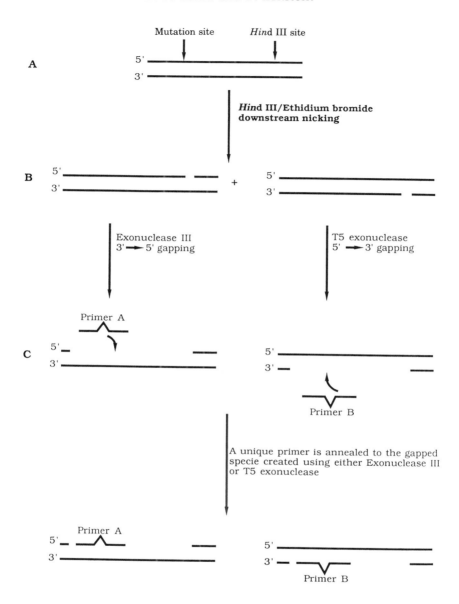

Figure 2. Representation of the two possibilities for 'productively gapping' nicked plasmid DNA using either the 3'→5' specific activity of exonuclease III or 5'→3' activity of T5 exonuclease. The lettering corresponds to the same steps as in *Figure 1*. The DNA species obtained after a nicking reaction (B) can be digested in the gapping reaction by either a 3'→5' exonuclease or a 5'→3' exonuclease depending on which mismatch primer has been chosen for the next step (C). After digestion (C) a unique mutant oligonucleotide (primer A or primer B) is required to anneal to the single-stranded DNA region.

91

Protocol 6. *continued*

2. Incubate at 70°C for 10 min and then place the tube in a 56°C heating block.

3. Allow the block to slowly cool to 37°C over 30 min.

---

a The amount of primer given in this protocol assumes an 18-mer oligonucleotide and 20 μg of starting pUC19 plasmid DNA. For significantly longer primers or larger vectors the amounts should be adjusted accordingly.

2.2.4 Preparation of the covalently closed circular mutant heteroduplex

Polymerization of the primed gapped plasmid is carried out using the Klenow fragment of DNA polymerase I at 16°C (see *Protocol 7*). It is very important that the reaction be carried out at a low temperature so that the enzyme is less likely to strand displace the mutant oligonucleotide (for more details of strand displacement see discussion in Chapter 3). Typically, after overnight polymerization a new band can be observed by agarose gel electrophoresis, which usually migrates faster than supercoiled plasmid.

Protocol 7. Preparation of the mutant heteroduplex

1. Add the following reagents to the sample solution from *Protocol 6*.
   6 μl 10 mM dATP
   6 μl 10 mM dCTP
   6 μl 10 mM dTTP
   6 μl 10 mM dGTPαS*a*
   6 μl 100 mM dithiothreitol
   12 μl 100 mM MgCl$_2$
   25 μl 10 mM ATP
   10 μl 500 mM Tris–HCl (pH 8)
   2 μl Klenow polymerase 5 units/μl
   8 μl T4 ligase 2 units/μl

2. Add sterile distilled water to bring the volume to 210 μl.

3. Briefly vortex the tube to mix the contents and spin to collect the solution in the bottom.

4. Incubate at 16°C overnight.

5. Remove 6 μl for agarose gel electrophoresis.

---

a The choice of nucleoside phosphorothioate (obtained from Amersham) used in the polymerization is dependent upon the restriction endonuclease used in the next step.

2.3 Creation of the mutant homoduplex

2.3.1 Mutant strand-selection by reaction with a restriction endonuclease inhibited by phosphorothioates

Protocol 8 gives the conditions for the nicking/linearization reaction using *Pst*I. This restriction endonuclease has a site which is adjacent to the *Hind*III site in many plasmid vector polylinkers. Therefore it can be used as the enzyme to give strand-selectivity against the wild-type sequence in most situations. This selectivity is based upon the inability of the enzyme to cleave a DNA strand which had been synthesized using dGTPαS substituted for dGTP in the polymerization reaction. However, there are a number of other restriction endonucleases which can also be used for this reaction. *Table 1* lists

Table 1. List of restriction endonucleases which are unable to linearize phosphorothioate containing DNA

Enzyme	DNA[a]	Polymerization using	Reference[b]
*Ava*I	M13mp2	dCTPαS	11
	øX174	dTTPαS	11
*Ava*II	M13mp2	dGTPαS	11
*Bam*HI	M13mp18	dATPαS/dGTPαS	13
*Ban*II	M13mp2	dCTPαS/dNTPαS[c]	11
*Eco*RI	M13mp18	dATPαS/dGTPαS	13
*Eco*RV	M13mp18	dATPαS	13
*Fsp*I	M13mp18	dGTPαS	7
*Hind*II	M13mp2	dGTPαS	7,11
	M13mp9		
	M13mp18		
*Hind*III	M13mp18	dATPαS/dTTPαS	13
*Nci*I	M13mp2	dCTPαS	7,11
	M13mp18		
*Pst*I	M13mp9	dGTPαS	7,10,11
	M13mp18		
	pUC19		
*Pvu*I	M13mp2	dCTPαS	11
*Pvu*II	M13mp18	dCTPαS/dGTPαS	13
*Sac*I	M13mp18	dCTPαS/dGTPαS	13
*Sma*I	M13mp18	dGTPαS[d]	12

[a] The initial nicking conditions were determined using the DNA vectors listed.

[b] It is recommended that all nicking reactions be carried out according to the buffer and incubation conditions given in the original reference.

[c] *Ban*II recognizes the sequence GPuGCPyC. Our unpublished results indicate that in order to inhibit this enzyme a phosphorothioate must be at the two positions designated by an asterisk in the following sequence 5'-GPuGCPy*C*-3'. In M13mp2 DNA this is accomplished by polymerization with dCTPαS. However, M13mp18 has a G 3' to the recognition sequence in the (−)strand and therefore the polymerization reaction requires both dCTPαS and dGTPαS.

[d] Nicking using *Sma*I requires the presence of dGMPS at the site of cleavage and 40 μg/ml ethidium bromide in the reaction (see ref. 12).

a number of restriction endonucleases that are unable to cleave DNA in which the normal dNMP linkages are replaced with dNMPαS (7, 11–13). More recent studies have revealed that DNA polymerized using two dNTPαS along with two dNTPs is stable against the catalytic activity of several other restriction endonucleases (13).

Protocol 8. *Pst*I nicking and linearization reaction

1. After extraction and spin-dialysis (see *Protocol 2*) bring volume to 85 μl with distilled sterile water.

2. Add:

70 units *Pst*I (20 units/μl)
10 μl 10 × *Pst*I buffer[a]

3. Briefly vortex the tube to mix the contents and spin to collect the solution in the bottom.

4. Incubate at 37°C for 80 min.

5. Remove 3 μl for agarose gel electrophoresis.

6. Repeat extraction and spin-dialysis (*Protocol 2*)

[a] 10 × *Pst*I buffer contains 100 mM Tris–HCl (pH 7.4), 100 mM MgCl$_2$, 1 M NaCl.

In order to obtain high mutational efficiencies it is critical that this reaction be carried out to completion since it is essential that all the wild-type species present in the reaction be hydrolysed during this step. As depicted in *Figure 1*, over 50% of the DNA population in solution do not contain phosphorothioates at the site of cleavage of *Pst*I. Therefore, analysis by agarose gel electrophoresis after reaction will reveal a large amount of linear plasmid along with some nicked product. There should be no evidence of any remaining covalently closed circular DNA.

2.3.2 Removal of the wild-type sequence by exonuclease digestion

The purpose of the second exonuclease digestion is to remove the wild-type sequence that is not base-paired with the mutant strand. We have used three different enzymes for this step without significant variation in the percent mutational efficiency. Conditions are given for digestion by T7 exonuclease in *Protocol 9* which can be used irrespective of the location of the nicking site in relation to the mismatch oligonucleotide. This enzyme is processive and in our hands entire plasmids have been gapped without significant degradation of the DNA. It is highly recommended that the buffer used for this gapping reaction be prepared immediately before use, otherwise severe degradation of the DNA can occur.

Protocol 9. Gapping of the nicked mutant heteroduplex

1. Adjust the volume of the solution from *Protocol 8* to 90 μl with water and add:

   10 μl of freshly prepared 10 × E9 buffer (see *Protocol 4*)
   1 μl T7 exonuclease 100 units/μl (US Biochemicals)

2. Briefly vortex the tube to mix the contents and spin to collect the solution in the bottom.

3. Incubate at 37 °C for 30 min.

4. Heat inactivate the enzyme by incubation at 70 °C for 10 min and place directly into a 37 °C heating block for 20 min.

5. Remove 10 μl for agarose gel electrophoresis.

If the non-processive enzyme exonuclease III is used for this step, it is important that the nick in the wild-type strand (as a result of the previous reaction) is 3' to the mutation. Otherwise, the enzyme would have to gap nearly the entire wild-type strand before digesting past the mutation. This can cause a decrease in mutational efficiency because exonuclease III is known to leave nearly 100 base-pairs remaining on the single-stranded template after completion of digestion. Therefore, the nick may never be gapped past the mismatch oligonucleotide. Alternatively, if the nick is 5' to the site of mutation, the 5'→3' activity of T5 exonuclease could be used to digest the wild-type strand.

It is interesting to note that the linear wild-type DNA resulting from the *Pst*I reaction will have 4 base-pair 3'-protrusions. Exonuclease III is unable to hydrolyse such DNA (19). However, T7 exonuclease or T5 exonuclease rapidly digests these substrates, as has been observed by agarose gel electrophoresis.

2.3.3 Repolymerization using the mutant strand as template

Following the second gapping reaction, the exonuclease is inactivated by heating at 70 °C for 10 min. The sample is then placed at 37 °C to anneal the small amount of wild-type strand remaining after digestion to the mutant strand. This serves as the substrate for repolymerization with DNA polymerase I using the mutant strand as template (see *Protocol 10*). After the reaction is complete an aliquot is withdrawn from the reaction for gel analysis. In some cases it is difficult to visualize any religated mutant homoduplex product on the gel. However, enough DNA is usually present for multiple transformations if required. The repolymerization step can be omitted at the cost of some loss of mutational efficiency (10).

Protocol 10. Preparation of the mutant homoduplex

1. Add the following reagents to the solution containing the gapped DNA from *Protocol 9*:

   6 μl 10 mM dATP
   6 μl 10 mM dCTP
   6 μl 10 mM dGTP
   6 μl 10 mM dTTP
   6 μl 100 mM dithiothreitol
   12 μl 100 mM MgCl$_2$
   25 μl 10 mM ATP
   20 μl 500 mM Tris–HCl (pH 8)
   1 μl DNA polymerase I 10 unit/μl
   8 μl T4 DNA ligase 2 units/μl

2. Add sterile distilled water to bring the volume to 220 μl.

3. Briefly vortex the tube to mix the contents and spin to collect the solution in the bottom.

4. Incubate at 16°C overnight or at 37°C for 3 h.

5. Remove 14 μl for agarose gel electrophoresis.

6. Remove 2 μl for transformation (*Protocol 12*).

2.4 Transformation of the mutant DNA

A procedure taken from Chung *et al.* (20) for the preparation of competent cells is given in *Protocol 11*. The *E. coli* cell lines TG-1 and SMH50 have yielded high transformational efficiencies but it is worth noting that this method can also be used with a number of other *E. coli* strains. We have had very reproducible results using this method even when the cells were stored at −80°C for several months. Transformations can be carried out using the procedure given in *Protocol 12*.

Protocol 11. Preparation of competent bacterial cells [a]

1. Add 3 ml of an overnight culture of TG-1 or SMH50 cells to 100 ml of LB media [b] in a 250-ml Erlenmeyer flask.

2. Incubate with shaking until A$_{600}$ equals 0.3–0.4.

3. The cells are harvested by centrifugation at 1000 g for 10 min at 4°C.

4. Add 10 ml of TSS buffer. [c]

5. Aliquot the cells into sterile polypropylene tubes and freeze using liquid nitrogen.

6. The cells are stored frozen at $-80\,^\circ$C.

[a] According to the procedure of Chung *et al.* (20).
[b] See *Protocol 1*.
[c] TSS buffer consists of LB media (pH 6.5) containing $20\,\text{mM}$ $MgCl_2$, 10% (w/v) PEG and 5% (v/v) DMSO.

Protocol 12. Transformation of competent cells

1. Place a microcentrifuge tube containing $100\,\mu$l of competent cells prepared in *Protocol 11* on ice.

2. Add $2\,\mu$l of repolymerized DNA (*Protocol 10*), mix gently and incubate on ice for 30 min.

3. Take 2, 10, and $80\,\mu$l of the transformed cells and spread on to agar plates containing the appropriate antibiotic for selection.

4. On another plate spread $10\,\mu$l of competent cells that have not come into contact with any DNA.

5. Incubate the plates overnight at $30\,^\circ$C and pick 2–5 colonies for DNA characterization.

---

3. Monitoring the procedure by agarose gel electrophoresis

The procedures given in this chapter normally require that an aliquot of the DNA be taken after each enzymatic reaction. These samples should be run on 1–2% agarose gels (depending on the size of the vector being used) which contain ethidium bromide and β-mercaptoethanol in the running buffer (21; see Chapter 3). For the novice it is sometimes difficult to discern whether a band corresponds to an expected species of DNA. For this reason we recommend that a linear marker of the starting plasmid be prepared by reaction with a restriction endonuclease which cleaves the DNA only once. This marker should be run along with a sample of the starting material with every analysis.

One of the causes of decreased mutational efficiencies could be the presence of a large amount of concatemeric DNA (the presence of these species also makes gel analysis sometimes very difficult). If large amounts of concatemers are visible by gel analysis we recommend altering the growth conditions of the host *E. coli* strain. In particular, we have found that amplification of the plasmid pUC19 by chloramphenicol greatly improves the quality of the DNA. Alternatively, we have also found that concatemeric DNA can be removed by denaturing-anion-exchange HPLC, using the Nucleogen column supplied by Macherey-Nagel.

4. Discussion

At several stages of the procedure the DNA is phenol-extracted and spin-dialysed using a Centricon-30 microconcentrator. After the first nicking reaction phenol-extraction removes both the ethidium bromide from solution and the large amount of restriction endonuclease. Subsequent spin-dialysis removes traces of dye or organic solvent remaining in solution. It is also important to vigorously vortex the extraction solution after the *Pst*I nicking step because this enzyme has been found to bind tightly to the nicked phosphorothioate-containing DNA (7). If the restriction endonuclease is not removed from solution subsequent exonuclease digestion of the wild-type strand could be inhibited, which would lead to decreased mutational efficiencies.

The procedure given in this chapter facilitates mutagenesis using double-stranded DNA because it circumvents the need for subcloning into single-stranded DNA vectors. The annealing step is the most critical phase of the plasmid mutagenesis protocol. At this stage the wild-type template is already primed for repolymerization since only a small stretch of single-stranded DNA was created in the previous step. It is important to make sure that the molar ratio of primer to template is 2:1 or slightly higher because any unprimed template will also be polymerized and it will not carry the desired change. Care should also be taken that the Klenow enzyme does not strand displace the mutant primer during this step. Therefore, the polymerization reaction is carried out at 16°C in order to decrease the likelihood of this event.

There are other advantages of performing phosphorothioate mutagenesis on double-stranded DNA. Since most of the DNA remains in the double-stranded form during the annealing step, the mutant oligonucleotide has a small chance of mispriming to unwanted regions of the template. In addition, because the gapped region is small the amount of phosphorothioates incorporated into the vector during repolymerization is limited (*Figure 1, C,D*). The small content of phosphorothioates ensures good transformability of the DNA; thus, the yield of colonies is high. As a result it is possible to transform the mutated DNA after the second exonuclease digestion. However, it is recommended that the repolymerization step be carried out since this further increases the mutational efficiency by approximately 20% (10).

Acknowledgements

The authors wish to express their thanks to J. Sayers and P. Padmanabhan for critical reading of the manuscript. We are also indebted to A. Fahrenholz and A. Wendler for expert technical assistance.

References

1. Knowles, J. R. (1987). *Science* **236,** 1252.
2. Gerlt, J. A. (1987). *Chemical Reviews*, **87,** 1079.
3. Shaw, W. V. (1987). *Biochemical Journal*, **246,** 1.
4. Kunkel, T. A., Roberts, J. D., and Zakour, R. A. (1987). In *Methods in Enzymology* (ed. R. Wu and L. Grossman), Vol. 154, pp. 367–82.
5. Kramer, W. and Fritz, H.-J. (1987). In *Methods in Enzymology* (ed. R. Wu and L. Grossman), Vol. 154, pp. 350–67.
6. Carter, P. (1987). *Methods in Enzymology* (ed. R. Wu and L. Grossman), Vol. 154, pp. 382–403.
7. Nakamaye, K. K. and Eckstein, F. (1986). *Nucleic Acids Research*, **14,** 9679.
8. Messing, J. (1983). In *Methods in Enzymology* (ed. R. Wu, L. Grossman and K. Moldave), Vol. 101, pp. 20–78.
9. Vieira, J. and Messing, J. (1987). In *Methods in Enzymology* (ed. R. Wu and L. Grossman), Vol. 153, pp. 3–11.
10. Olsen, D. B. and Eckstein, F. (1990). *Proceedings of the National Academy of Sciences of the USA*, **87,** 1451.
11. Taylor, J., Schmidt, W., Cosstick, R., Okruszek, A., and Eckstein, F. (1985). *Nucleic Acids Research*, **13,** 8749.
12. Sayers, J. R., Schmidt, W., Wendler, A., and Eckstein, F. (1988). *Nucleic Acids Research*, **16,** 803.
13. Sayers, J. R., Olsen, D. B., and Eckstein, F. (1989). *Nucleic Acids Research*, **17,** 9495.
14. Dalbadie-McFarland, G., Cohen, L. W., Riggs, A. D., Morin, C., Itakura, K., and Richards, J. H. (1982). *Proceedings of the National Academy of Sciences of the USA*, **79,** 6409.
15. Parker, R. C., Watson, R. M., and Vinograd, J. (1977). *Proceedings of the National Academy of Sciences of the USA*, **74,** 851.
16. Rawlins, D. R. and Muzyczka, N. (1980). *Journal of Virology*, **36,** 611.
17. Österlund, M., Luthman, S., Nilsson, S. V., and Magnusson, G. (1982). *Gene*, **20,** 121.
18. Shortle, D. and Nathans, D. (1978). *Proceedings of the National Academy of Sciences of the USA*, **75,** 2170.
19. Henikoff, S. (1984). *Gene*, **28,** 351.
20. Chung, C. T., Niemela, S. L., and Miller, R. H. (1989). *Proceedings of the National Academy of Sciences of the USA*, **86,** 2172.
21. Taylor, J. W., Ott, J., and Eckstein, F. (1985). *Nucleic Acids Research*, **13,** 8765.
22. Miller, H. (1987). *Methods in Enzymology* (ed. S. L. Berger and A. R. Kimmel), Vol. 152, pp. 145–70.
23. Woodhead, J. L., Bhave, N., and Malcolm, A. D. B. (1981). *European Journal of Biochemistry*, **115,** 293.
24. Maniatis, T., Fritch, E. F., and Sambrook, K. (ed.) (1982). *Molecular Cloning: A Laboratory Manual*, pp. 458–60. Cold Spring Harbor Laboratory, Cold Spring Harbor, New York.

6

Linker scanning mutagenesis

BRUNO LUCKOW and GÜNTHER SCHÜTZ

1. Introduction

1.1 Definition of linker scanning mutagenesis

Techniques for *in vitro* mutagenesis of DNA play a major role in molecular biology. They are powerful tools in dissecting the relationship between structure and function of a given gene. A typical analysis of gene function proceeds in several steps with an increasing degree of resolution. Deletion mutations are usually analysed first; once a functionally important region has been delimited by deletion mapping, this site can be characterized in more detail by linker scanning mutagenesis and by single point mutations introduced by methods described elsewhere in this book.

In linker scanning (LS) mutations a short segment of the DNA of interest is replaced by a new synthetic sequence without inserting or deleting base-pairs, thereby introducing a cluster of point mutations at the replacement site. Generation of a series of mutations in which the replacement site is systematically moved, allows scanning of the DNA of interest to identify regions which are critical for the activity of a gene product or in regulating gene expression. LS mutations give a relatively high degree of resolution, while the number of mutations required to perform a systematic analysis of a region of interest is reduced, compared to single base substitutions. In contrast to deletions, LS mutations do not alter the original distances between different DNA sequences. This makes the unequivocal interpretation of experimental results much easier.

Linker scanning mutagenesis is therefore an excellent method for detailed analysis of complex regulatory regions such as eukaryotic promoters, but this method can also be used for characterization of DNA regions coding for a protein (1–15).

1.2 Different strategies for the construction of linker scanning mutations

The technique of linker scanning mutagenesis was developed in 1982 by McKnight and Kingsbury (1), who constructed a representative series of LS mutations of the Herpes simplex virus thymidine kinase (HSV-tk) promoter.

101

First, they generated a series of 5′-deletions and a series of 3′-deletions of the HSV-tk promoter. Synthetic linkers with a length of 10 bp were subsequently ligated to the newly created ends. The exact deletion end-points were determined by sequencing thus allowing the identification of matching pairs consisting of 5′- and 3′-deletions with deletion end-points just 10 bp apart. The 5′- and 3′-deletion of such a matching pair could be joined *via* the identical linker sequence, giving rise to a LS mutation in which 10 base-pairs of the original sequence were exactly substituted by the linker sequence.

In 1985 Haltiner and colleagues published some modifications of the original McKnight procedure (16). In summary, these modifications simplified the ligation of linkers as well as the sequence determination of mutations, although the principle to generate LS mutations by recombining matching 5′- and 3′-deletions was retained. This strategy has one major drawback however; a large number of mutations has to be sequenced in order to find a few matching pairs. Furthermore, each LS mutation has to be constructed separately by the ligation of matching deletion fragments.

In an attempt to overcome these inherent difficulties, we have developed a completely different approach for the construction of LS mutations (17). First, plasmids containing the DNA of interest are linearized by a randomly placed cut. Then a few base-pairs are removed from the ends of these plasmids. After the ligation of linkers the plasmids are recircularized. Subsequently, all plasmids are isolated which have the linker inserted anywhere within the DNA of interest. Correct LS mutations are identified on the basis of the topoisomer pattern and finally verified by determination of the DNA sequence. A schematic outline of this new strategy for linker scanning mutagenesis is shown in *Figure 1*. Detailed explanations are given in Section 2.

1.3 Alternative scanning mutagenesis procedures

In the course of the last few years, several alternative procedures to linker scanning mutagenesis have been described. The oligo-scanning mutagenesis procedure (18) represents a variation of the oligonucleotide-directed mutagenesis protocol (19) in which an extended mismatch primer can be used to introduce a cluster of point mutations in a single step. A specific oligonucleotide of at least 30 bases is required to mutate a block of 10 bp. Mutations are recovered with a frequency in the range of 1–7%. Due to this relatively

---

Figure 1. Scheme for linker scanning mutagenesis. All steps involved in generating linker scanning mutations are listed. In addition the different structures of the 'starting plasmid' at various stages of the mutagenesis procedure are shown. The DNA fragment to be mutated (DNA of interest) is represented by a thick black line, whereas the fragment conferring resistance to kanamycin (Km^R) is represented by a thick grey line. The letters A and B symbolize different unique restriction sites flanking the DNA of interest. The letter X indicates other restriction sites present on the linker used for linker scanning mutagenesis and at the borders of the Km^R fragment. For further explanations, see text.

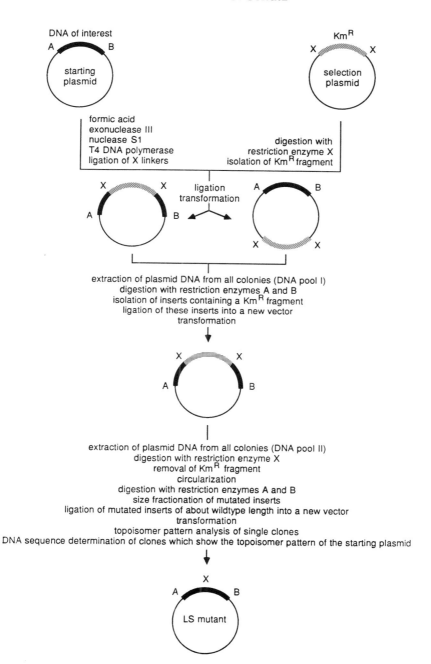

low efficiency, the desired mutation has to be identified on the basis of differential hybridization signals.

The TAB (Two Amino acid Barany) linker mutagenesis procedure (20) uses single-stranded hexameric linkers to create two-codon insertions at restriction sites. The dependence on pre-existing restriction sites represents a significant limitation of this method. TAB linker mutations are therefore useful to analyse larger DNA regions coding for a protein, but they are less suited for studying sequences regulating gene expression.

2. Construction of linker scanning mutations

Section 2 is devoted to practical aspects of linker scanning mutagenesis. Detailed methods are given in *Protocols 1–21*. A schematic outline of the mutagenesis procedure is presented in *Figure 1*.

2.1 Apurination of plasmid DNA by formic acid

The procedure for linker scanning mutagenesis as described in this chapter starts out with a formic acid treatment of the so-called 'starting plasmid'. This plasmid consists of vector sequences and the DNA fragment which is to be mutated, the so-called 'DNA of interest'. The incubation of the 'starting plasmid' with formic acid leads to an acid-catalysed partial apurination of the supercoiled plasmid DNA (21). The extent of apurination is determined by the incubation time and the temperature. The aim of this reaction is the generation of a single, randomly placed apurinic site within the 'starting plasmid'. The DNA will be cleaved at a later stage exactly at the position of the apurinic site. To keep the portion of plasmids with more than one apurinic site low, the reaction is performed so that only 30–50% of all plasmids contain apurinic sites. Unfortunately, the reaction cannot be checked directly since apurinated plasmid DNA has the same covalently closed circular configuration as untreated plasmid DNA (see *Figure 2*, lanes 2 and 4). However, the reaction conditions proposed in *Protocol 1* should yield satisfying results.

Protocol 1. Apurination of plasmid DNA by formic acid

1. Incubate 50 μg 'starting plasmid' (supercoiled form) dissolved in 100 μl double distilled water, prepared according to *Protocol 10*, for 15 min at 15 °C.

2. Dilute 2 μl pure formic acid (98–100%) with 98 μl double distilled water and incubate this dilution (pH = 2) for 15 min at 15 °C.

3. Add 10 μl of the formic acid dilution to the tube with the 'starting plasmid'. Mix by pipetting up and down. Incubate for 4 min at 15 °C.

4. Neutralize by the addition of 40 μl 1 M Tris–HCl (pH 7.5).

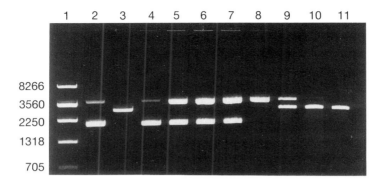

Figure 2. Intermediates from the construction of linker scanning mutations (I). Samples were taken at various stages of the linker scanning mutagenesis procedure and subsequently analysed on a 0.6% agarose gel. Lane 1, size standards; lane 2, 'starting plasmid' DNA, supercoiled form; lane 3, 'starting plasmid' DNA, linear form; lane 4, same DNA as shown in lane 2 after treatment with formic acid; lanes 5 to 7, same DNA as shown in lane 4 after treatment with exonuclease III (3 different time points); lane 8, isolated nicked and gapped circles from the DNAs shown in lanes 5 to 7; lanes 9 to 11, same DNA as shown in lane 8 after treatment with nuclease S1 (3 different time points). Reproduced with permission from ref. 17.

5. Add $15\,\mu$l 3 M sodium acetate (pH 9) and $400\,\mu$l ethanol. Precipitate the DNA for 15 min in a dry-ice/ethanol bath.

6. Collect the DNA by centrifugation ($12\,000\,g$, 15 min). Remove the supernatant.

7. Wash the pellet with $500\,\mu$l 80% ethanol, centrifuge briefly, discard the supernatant, dry the pellet in a SpeedVac for 1 to 2 min or simply in the air.

2.2 Generation of nicks and small gaps in apurinic DNA by exonuclease III

After the formic acid treatment, the 'starting plasmid' is incubated with exonuclease III. This enzyme possesses, in addition to its well-known $3' \rightarrow 5'$ exonuclease activity, an endonuclease activity specific for apurinic DNA (22). The incubation of the 'starting plasmid' with exonuclease III therefore leads to the generation of nicks and small single-stranded gaps of a few bases specifically at the apurinic sites. The incubation time is varied between 1 and 9 min to compensate for differences in reactivity of various sequences. By the use of a reaction buffer with Ca^{2+} ions instead of Mg^{2+} ions, the exonuclease activity is suppressed almost completely without affecting the AP endonuclease activity (22). This makes it possible to nick the DNA at all apurinic sites without generating simultaneously extended gaps. Since only plasmids with an apurinic site are nicked by exonuclease III, two different DNA species are visible on a gel after the reaction. An example is shown in *Figure 2*. The exact conditions for the exonuclease III reaction are given in *Protocol 2*.

Protocol 2. Generation of nicks and small gaps at apurinic sites by exonuclease III

1. Dissolve the mixture of apurinated and unreacted supercoils (*Protocol 1*, step 7) in 100 μl TE buffer (10 mM Tris–HCl (pH 7.5), 1 mM EDTA).

2. Add 60 μl double distilled water and 40 μl 5 × exonuclease III buffer.[a]

3. Pre-incubate for 15 min at 37°C.

4. Add 200 units (= 2.5 μl) exonuclease III (Boehringer), mix and place at 37°C.

5. Take 66 μl aliquots after 1 min, 3 min, and 9 min.

6. Stop the reactions immediately by the addition of 1 μl 0.5 M EDTA and a phenol/chloroform (1:1, v/v) extraction.

7. Separate the phases by a brief centrifugation step (3 min, 12 000 g).

8. Pool the three aqueous phases and extract with chloroform/isoamyl-alcohol (24:1, v/v).

9. Extract with diethylether saturated with water.

10. Precipitate the DNA by adding 20 μl 3 M sodium acetate (pH 9) and 500 μl ethanol.

11. Incubate for 15 min in a dry-ice/ethanol bath.

12. Centrifuge for 15 min at 12 000 g. Discard the supernatant.

13. Wash the pellet with 500 μl 80% ethanol. Centrifuge 5 min at 12 000 g, discard the supernatant, dry the pellet (e.g. in a SpeedVac for 1–2 min).

[a] 5x exonuclease buffer is 330 mM Tris–HCl (pH 8.0), 625 mM NaCl, 25 mM $CaCl_2$, 50 mM DTT.

2.3 Linearization of nicked or gapped plasmid DNA by nuclease S1

After the exonuclease III reaction, the single-strand-specific nuclease S1 is used to linearize the nicked and gapped plasmid DNA (23). To avoid problems resulting from a S1 hypersensitive site potentially present in the 'starting plasmid', the remaining supercoiled DNA is removed before the S1 reaction is performed. We recommend separating the nicked/gapped circles from the remaining supercoils by a CsCl gradient rather than by an agarose gel to circumvent potential problems caused by the inhibition of T4 DNA polymerase in the subsequent step. If available, a table-top ultracentrifuge should be used, since it requires less time and DNA to perform the separation (see *Protocol 3*). There is, however, no absolute requirement for such a centrifuge. Increasing the amount of 'starting plasmid' from $50\,\mu g$ to $200\,\mu g$ allows one to perform the separation overnight in a standard ultracentrifuge, using a Beckman VTi65 or equivalent rotor.

Protocol 3. Separation of nicked and gapped circles from supercoiled plasmid DNA

1. Dissolve the mixture of nicked and gapped circles and remaining super-coils (*Protocol 2*, step 13) in $180\,\mu l$ TE buffer.

2. Add $20\,\mu l$ ethidium bromide ($10\,mg/ml$ in H_2O).

3. Add $900\,\mu l$ of a $7.1\,M$ CsCl solution.

4. Transfer the solution to a quick seal tube for the Beckman TLV-100 rotor using a Pasteur pipette.

5. Fill up with 1 ml of a $3.8\,M$ CsCl solution. Avoid mixing of the two CsCl layers.

6. Seal the tube by heat and centrifuge it for $2.5\,h$ at $100\,k.r.p.m.$ and $20\,°C$ in the vertical rotor TLV-100 in a Beckman table-top ultracentrifuge TL-100. (Alternatively, perform the separation in a standard ultracentrifuge overnight, see text.)

7. Collect the upper of the two bands visible after the run. It will contain the nicked and gapped circles.

8. Remove the ethidium bromide by several extractions with isopropanol saturated with CsCl.

9. Dilute by adding 4 vol. of double distilled water.

10. Precipitate the DNA in a dry-ice/ethanol bath for $15\,min$ after addition of $2.5\,vol.$ ethanol.

11. Centrifuge for $15\,min$ at $12\,000\,g$.

12. Wash the pellet with 80% ethanol, dry the pellet in the air.

Subsequent to their isolation, the nicked and gapped plasmids are cleaved by nuclease S1 exactly at the position of the nicks and gaps. The incubation time is varied between 1 and $15\,min$ for the same reason as mentioned in the case of the exonuclease III reaction. The pH of the S1 reaction buffer is increased from 4.5 to 5.7 to prevent any additional apurination of the nicked/gapped circles. As a consequence, the enzymatic activity is considerably decreased. The nuclease S1 reaction leads not only to a linearization of the nicked or gapped plasmids but also to a slight shortening of the linear plasmids, which is most likely caused by breathing of the ends. Using the conditions given in *Protocol 4*, the enzyme removes between 0 and $30\,bp$. A gel with plasmid DNA before and after the S1 reaction is shown in *Figure 2*.

Protocol 4. Linearization of nicked/gapped circles by nuclease S1

1. Dissolve the mixture of nicked and gapped circles (*Protocol 3*, step 12) in 105 μl double distilled water.
2. Add 12 μl 10 × S1 buffer^a and pre-incubate for 15 min at 37 °C.
3. Start the reaction by adding 1200 units (= 3 μl) nuclease S1 (Sigma), mix and incubate at 37 °C.
4. Take 40 μl aliquots after 1 min, 4 min, and 15 min.
5. Stop the reactions immediately by the addition of 1 μl 0.5 M EDTA and a phenol/chloroform (1:1, v/v) extraction.
6. Separate the phases by brief centrifugation (3 min, 12 000 g).
7. Pool the three aqueous phases and extract with chloroform/isoamyl alcohol (24:1, v/v).
8. Extract with diethylether saturated with water.
9. Precipitate the DNA with 12 μl 3 M sodium acetate (pH 9) and 300 μl ethanol for 15 min in a dry-ice/ethanol bath.
10. Centrifuge for 15 min at 12 000 g. Discard the supernatant.
11. Wash the pellet with 500 μl 80% ethanol. Centrifuge 5 min at 12 000 g, discard the supernatant, dry the pellet (e.g. in a SpeedVac for 1–2 min).

^{a} 10 × S1 buffer is 500 mM sodium acetate (pH 5.7), 2 M NaCl, 10 mM ZnSO$_4$, 5% glycerol.

2.4 Ligation of linkers

A significant portion of the plasmids linearized by nuclease S1 has overhanging ends. Such plasmids are unsuitable substrates for the following linker ligation reaction. Therefore, the plasmids are treated first with T4 DNA polymerase, according to *Protocol 5*, to maximize the number of molecules with blunt ends.

Protocol 5. Generation of blunt ended plasmids by T4 DNA polymerase.

1. Dissolve the linearized and slightly shortened plasmids (*Protocol 4*, step 11) in 79 μl TE buffer.
2. Add 10 μl 10 × T4 DNA polymerase buffer.^{a}
3. Add 5 units (= 1 μl) T4 DNA polymerase (BRL), mix and incubate for 3 min at room temperature.
4. Add 10 μl of a 1 mM solution of all four deoxynucleoside triphosphates.
5. Incubate for 30 min at 37 °C.

6. Stop the reaction by addition of $2\,\mu l$ $0.5\,M$ EDTA and a phenol/chloroform extraction.

7. Extract with chloroform/isoamyl alcohol.

8. Extract with diethylether saturated with water.

9. Add $50\,\mu l$ of a $7.5\,M$ ammonium acetate solution.

10. Precipitate the DNA with $400\,\mu l$ ethanol for 15 min in a dry-ice/ethanol bath.

11. Centrifuge for 15 min at $12\,000\,g$.

12. Wash the pellet twice with $500\,\mu l$ 80% ethanol. Centrifuge 5 min at $12\,000\,g$, discard the supernatant, dry the pellet (e.g. in a SpeedVac for 1–2 min).

^{a} 10 ×T4 DNA polymerase buffer is 200 mM Tris–HCl (pH 7.6), 100 mM MgCl$_2$, 10 mM DTT.

Unphosphorylated linkers are kinased, as described in *Protocol 6*, in the presence of $[\gamma\text{-}^{32}P]ATP$ to allow the monitoring of the ligation reaction.

Protocol 6. Phosphorylation of linkers by T4 polynucleotide kinase

1. Dry down $50\,\mu Ci$ $[\gamma\text{-}^{32}P]ATP$ (5000 Ci/mmol) in a siliconized microcentrifuge tube using a SpeedVac.

2. Add 1 nmol linkers, $4\,\mu l$ 10 × kinase buffer^a and double distilled water to $39\,\mu l$.

3. Add 10 units ($= 1\,\mu l$) T4 polynucleotide kinase (Pharmacia).

4. Incubate for 30 min at 37°C.

5. Add $1\,\mu l$ 10 × kinase buffer, $5\,\mu l$ water, $4\,\mu l$ of a 1 mM solution of cold ATP and another 10 units kinase.

6. Incubate for 30 min at 37°C.

7. Heat inactivate the enzyme for 10 min at 70°C.

8. Store the kinased linkers at −20°C.

^{a} 10 × kinase buffer is 660 mM Tris–HCl (pH 7.6), 100 mM MgCl$_2$, 50 mM DTT.

These synthetic linkers are then ligated to the ends of the slightly shortened plasmids (see *Protocol 7*).

Protocol 7. Ligation of linkers by T4 DNA ligase

1. Dissolve the randomly linearized, slightly shortened blunt-ended plasmids (*Protocol 5*, step 12) in $100\,\mu l$ TE buffer.

2. Transfer an aliquot of this solution containing $10\,\mu g$ of DNA (approximately 10 pmol 5'-ends) to a new microcentrifuge tube.

Protocol 7. *continued*

3. Add 20 μl 10 × ligase buffer,^{a} 10 μl of a 10 mM ATP solution, 1000 pmol kinased linkers (*Protocol 6*, step 8) and double distilled water to give a volume of 180 μl.

4. Start the reaction by adding 8000 units (= 20 μl) T4 DNA ligase (New England Biolabs).

5. Mix and incubate overnight at 15 °C.

6. Heat inactivate the enzyme for 10 min at 70 °C.

7. Take a 2 μl aliquot to check the efficiency of the linker ligation at a later step.

8. Add 50 μl of the appropriate 10 × buffer for the restriction enzyme X, 200 μl double distilled water and 500 units (= 50 μl) of restriction enzyme X, which cleaves the linkers.

9. Incubate for several hours at the appropriate temperature.

10. Take a 5 μl aliquot after 2.5 h to check the completeness of digestion with restriction enzyme X.

11. Load the two aliquots taken at steps 7 and 10 on a non-denaturing 10% polyacrylamide gel (20 × 40 × 0.1 cm) run with 750 V in 0.5 × TBE buffer (45 mM Tris-borate, 45 mM boric acid, 1 mM EDTA) until the Bromophenol Blue of the loading buffer has migrated 10 cm.

12. Cover the gel with Saran Wrap and expose for 2 h at room temperature.

13. Develop the film. A successful linker ligation is indicated by the appearance of a 'linker ladder' on the autoradiograph. It should disappear upon digestion with restriction enzyme X.

14. Stop the restriction enzyme digestion by the addition of 10 μl 0.5 M EDTA and a phenol/chloroform extraction.

15. Extract with chloroform/isoamyl alcohol.

16. Extract with diethylether saturated with water.

17. Precipitate the DNA by adding 0.5 vol. 7.5 M ammonium acetate and 2.5 vol. ethanol. Incubate for 15 min in a dry-ice/ethanol bath.

18. Centrifuge for 15 min at 12 000 g. Discard the supernatant.

19. Resuspend the pellet in 200 μl TE. Repeat steps 17 and 18.

20. Wash the pellet with 500 μl 80% ethanol. Centrifuge 5 min at 12 000 g. Discard the supernatant. Repeat this step.

21. Dry the pellet for 1 to 2 min in a SpeedVac and dissolve it in 50 μl TE buffer.

22. Check a 1 μl aliquot on a 0.6% agarose minigel.

^{a} 10 × ligase buffer is 500 mM Tris–HCl (pH 7.8), 100 mM MgCl$_2$ 200 mM DTT.

In the ideal case, the original length of the 'starting plasmid' is restored by the addition of the linkers whereby correct LS mutations are generated. In practice, however, this is only true for a low percentage of all plasmids and several enrichment steps are necessary to isolate plasmids carrying correct LS mutations within the DNA of interest.

2.5 Generation of DNA pool I

Since the ligation of linkers is usually not a very efficient process, plasmids with linkers attached to both ends must be selected. Otherwise, a high background of plasmids without a linker will be obtained in the following transformation step. The selection makes use of a DNA fragment, which contains a drug resistance gene and which carries at both ends the same linkers as the 'starting plasmid'. The aminoglycoside 3'-phosphotransferase gene from transposon Tn903 (24) conferring resistance to kanamycin is well-suited for this purpose. The gene is isolated as a 1.4 kb *Hae*II fragment and inserted into a suitable vector after the ligation of the desired linkers thereby generating the so-called 'selection plasmid'. This plasmid serves subsequently as a source for the Km^R fragment. However, it is not always necessary to construct a 'selection plasmid'. The plasmid pBL2, which can be obtained from the authors, allows excision of the Km^R fragment either with *Bam*HI or *Bgl*II (17). From the plasmids of the pUC4-K series, which are available from Pharmacia, this fragment can be excised with *Apa*I, *Bam*HI, *Eco*RI, *Kpn*I, *Pst*I, *Sal*I, and *Sph*I.

A ligation reaction is set up containing equimolar amounts of linear 'starting plasmids' with attached linkers and the Km^R fragment as described in *Protocol 8*.

Protocol 8. Generation of DNA pool I

1. Pipette an aliquot containing 1 μg 'starting plasmids' with attached X linkers (*Protocol 7*, step 21) into a 1.5 ml microcentrifuge tube.

2. Add 1 μg of the kanamycin resistance (Km^R) fragment carrying X linkers at the ends, 5 μl 10 × ligase buffer (see *Protocol 7*), 3 μl 10 mM ATP, double distilled water to 49 μl and 400 units (= 1 μl) T4 DNA ligase (New England Biolabs).

3. Mix and incubate at least for 6 h at 15°C.

4. Use a 1 μl aliquot for the colony transformation procedure (*Protocol 9*).

5. Perform five large scale colony transformations with the entire ligation reaction (*Protocol 9* and Section 2.5).

6. Plate between 250 000 and 500 000 transformants on 20 large LB plates[a] (Nunc Petri dishes, 24 × 24 cm) containing ampicillin and kanamycin each at 100 μg/ml.

7. Incubate the plates upside down for 20 to 24 h at 37°C.

8. Wash the colonies off the plates using LB medium and an L-shaped glass rod.

Protocol 8. *continued*

9. Collect the bacteria by centrifugation for 10 min at 5000 g.
10. Extract the plasmid DNA ('DNA pool I') according to *Protocol 10*. The yield of bacteria from four large LB plates is roughly equivalent to that from a 1 litre liquid culture.

ᵃ LB agar is 10 g bactotryptone, 5 g bactoyeast extract, 10 g NaCl; adjust to pH 7.5 with NaOH, 15 g bactoagar per litre.

This ligation reaction is subsequently used to transform bacteria by a highly efficient transformation procedure. We recommend the colony transformation protocol developed by Hanahan (26) as transformation efficiencies obtained with this procedure are routinely in the range of 5×10^7 transformants per microgram of supercoiled pUC18 DNA, using the *E. coli* strain JM109. However, alternative protocols such as electroporation (27) might be suitable as well. For preparative purposes, the transformation procedure described in *Protocol 9* can be scaled up by a factor of 10. The bacteria are then dispersed in 2 ml TFB and 70 μl of DMSO or DTT are added. Such a large scale transformation is carried out in a 50 ml polypropylene tube and all incubation times remain the same with the exception of the heat shock, which has to be extended to 210 sec. Before plating, the cells are pelleted by centrifugation for 5 min at 800 g. The supernatant is discarded and the bacteria are resuspended in the desired volume of medium.

Protocol 9. Colony transformation

1. Streak the *E. coli* strain chosen for transformation (e.g. HB101, JM109) the day before on a SOB plate.*ᵃ* Start from a glycerol culture stored at −70 °C.
2. Incubate the plate overnight at 37 °C.
3. Take the plate out of the incubator and leave it at room temperature for at least half an hour.
4. Pipette 200 μl transformation buffer (TFB)*ᵇ* into a 17 × 100 mm polypropylene tube.
5. Pick a few colonies and disperse them by pipetting up and down until a thick suspension is produced. Avoid transferring pieces of agar as this inhibits transformation.
6. Incubate the tube on ice for 15 min.
7. Add 7 μl of a freshly thawed aliquot of dimethyl sulphoxide (DMSO). Mix by gentle vortexing or swirling for a few seconds.
8. Incubate the tube on ice for 5 min.

9. Add $7 \mu l$ DTT solution.[c] Mix as above.

10. Place the tube on ice for 10 min.

11. Add another $7 \mu l$ of DMSO. Mix as above.

12. Leave the tube on ice for 5 min.

13. Add the DNA in a volume of $\leqslant 10 \mu l$. Mix as above.

14. Incubate the tube for 30 min on ice.

15. Place the tube for 90 sec in a 42°C water bath and then chill on ice for 2 min.

16. Add $800 \mu l$ SOC medium.[d]

17. Shake for 30 to 60 min at 37°C.

18. Spread appropriate portions of the transformation mixture on plates containing suitable antibiotics to select transformants. The remainder can be stored overnight at 4°C without a significant decrease in transformation efficiency.

[a] SOB medium is 2% bactotryptone, 0.5% bactoyeast extract, 10 mM NaCl, 2.5 mM KCl, 10 mM $MgCl_2$, 10 mM $MgSO_4$; for plates add 1.5% agar.
[b] TFB is 100 mM KCl, 45 mM $MnCl_2$, 10 mM $CaCl_2$, 3 mM $HACoCl_3$, 10 mM K-Mes, (pH 6.2) $HACoCl_3$ is hexamine cobalt (III) trichloride, K-Mes is 2(N-morpholino)ethane sulphonic acid adjusted with KOH to pH 6.2.
[c] 2.2 M DTT in 100 mM potassium acetate (pH 7.5).
[d] SOB medium containing in addition 20 mM glucose.

Double selection for kanamycin and ampicillin resistance is performed to exclude plasmids containing the Km^R fragment within the β-lactamase gene of the 'starting plasmid'. This leads to an increase in the number of clones harbouring a plasmid which has the linker inserted in the DNA of interest. A representative library of mutated plasmids is established by plating an appropriate number of double resistant clones. The colonies are pooled and the plasmid DNA is extracted according to *Protocol 10* (25). This mixture of mutated 'starting plasmids' is called DNA pool I.

Protocol 10. Large scale preparation of plasmid DNA ('Maxiprep')

1. Grow the bacteria containing the plasmid of interest with vigorous shaking in a Fernbach flask for 1 to 2 days in 1 litre LB medium (see *Protocol 8*).

2. Harvest the cells by centrifugation (10 min, 5000 g, 4°C). Discard the supernatant. The pellet can either be used immediately or stored frozen at −20°C until needed.

3. Resuspend the bacteria in 28 ml STET buffer (50 mM Tris–HCl (pH 8.0), 50 mM EDTA, 5% Triton X-100, 8% sucrose) by pipetting up and down. Avoid excessive foaming.

Protocol 10. *continued*

4. Transfer the homogeneous suspension to a 100-ml Erlenmeyer flask.

5. Add 2.5 ml of a freshly prepared lysozyme solution (10 mg/ml in water).

6. Swirl and hold the flask in the open flame of a Bunsen burner until the liquid starts to boil. Shake the flask constantly. The suspension becomes very viscous during heating.

7. Immediately immerse the flask for 60 sec into a boiling-water bath.

8. Cool the flask for a few minutes in ice-cold water.

9. Transfer the lysate to a polycarbonate ultracentrifuge tube. Centrifuge at 60 000 g for 30 min and 4 °C. A clear supernatant and a compact pellet is obtained.

10. Transfer the supernatant to a 50 ml polypropylene tube and extract with 5 ml phenol/chloroform.

11. Centrifuge at 3000 g for 5 min at 4 °C.

12. Transfer the milky upper phase to a new 50 ml tube.

13. Add 0.8 vol. isopropanol, mix, leave at −20 °C for 1 h.

14. Centrifuge at 3000 g for 15 min at 4 °C. Discard the supernatant.

15. Wash the pellet with 5 ml 80% ethanol, centrifuge briefly (5 min, 4 °C, 3000 g), discard the supernatant.

16. Dry the pellet slightly for 15 to 30 min in a desiccator connected to a water suction pump or simply by blowing a stream of nitrogen over the pellet for a few minutes.

17. Dissolve the pellet in 3 ml TE by pipetting up and down.

18. Add 4.3 g solid CsCl.

19. Mix until the salt is completely dissolved.

20. Add 0.5 ml of a ethidium bromide solution (10 mg/ml in H_2O) Mix thoroughly. Purple flocks which form are complexes between ethidium bromide and proteins.

21. Centrifuge briefly (5 min, 3000 g).

22. Transfer the clear solution to a quick seal tube for a Beckman VTi65 or similar rotor.

23. Fill the tube either with water or a 50% CsCl solution (w/w) to the top so that the final weight is between 9.5 g and 9.7 g.

24. Seal the tube by heat.

25. Centrifuge at 265 000 g for 12 to 16 h at 20 °C. Two bands should be visible upon illumination with a long-wave (366 nm) UV lamp.

26. Collect the lower band containing closed circular plasmid DNA.

27. Transfer the plasmid DNA to a new tube for a Beckman VTi65 or similar rotor.

28. Fill up to the top with a 50% CsCl solution (w/w). The final weight of the tube should be around 9.5 g.

29. Seal the tube by heat.

30. Centrifuge again at 265 000 g for at least 5 h at 20 °C.

31. Collect the plasmid DNA.

32. Remove the ethidium bromide by several extractions with isopropanol saturated with CsCl.

33. Dilute with 4 vol. water before precipitating the DNA with 2 vol. ethanol overnight at −20 °C.

34. Collect the DNA by centrifugation (15 min, 12 000 g).

35. Discard the supernatant and wash the pellet once with 80% ethanol.

36. Dry the pellet in the air and dissolve it in 1 ml TE buffer.

37. Determine the yield of the isolated plasmid DNA by measuring the extinction of a diluted aliquot at 260 nm.

2.6 Generation of DNA pool II

Only a minor portion of the plasmids present in the DNA pool I will carry the linker within the DNA of interest. This small fraction of mutated inserts is isolated as described in *Protocol 11*. The isolation procedure exploits the fact that only mutated inserts contain the Km^R fragment. This makes it possible to separate wild-type inserts from mutated inserts simply by gel electrophoresis. Pool I DNA is digested with restriction enzymes A and B to excise all inserts, and the digestion products are separated on a gel.

Protocol 11. Isolation of mutated inserts from DNA pool I

1. Pipette 100 μg plasmid DNA (*Protocol 8*, step 12) into a 1.5 ml microcentrifuge tube.

2. Add 40 μl of the appropriate 10 × restriction enzyme buffer, 200 units each of restriction enzymes A and B (to excise the inserts; see *Figure 1*) and double distilled water to 400 μl.

3. Incubate for 2 h at the appropriate temperature.

4. Check the digestion on a 0.6% agarose minigel.

5. Stop the reaction by adding 10 μl 0.5 M EDTA.

6. Separate the digested pool I DNA directly on a 0.6% low melting point agarose gel run in 1 × Tris-acetate buffer (40 mM Tris-acetate (pH 7.8),

Protocol 11. *continued*

    2 mM EDTA) containing ethidium bromide (0.5 μg/ml). The running buffer should be circulated to avoid the generation of a pH gradient.

7. On a long-wave (366 nm) UV transilluminator excise the band corresponding to mutated inserts containing the Km^R fragment (see *Figure 3*, lane 3).

8. Determine the weight of the agarose slice.

9. Place it in a 15 ml polypropylene tube.

10. Add 3 M sodium acetate (pH 9) to a final concentration of 0.3 M.

11. Melt the agarose by placing the tube in a 70°C water bath.

12. Extract with 1 vol. phenol.

13. Separate the phases by a short centrifugation (5 min, 3000 g).

14. Transfer the aqueous phase to a new tube.

15. Extract in the same way with phenol/chloroform, chloroform/isoamyl alcohol and ether. If necessary, the volume can be decreased by several extractions with 2-butanol.

16. Precipitate the DNA by addition of 2.5 vol. ethanol and incubation in a dry-ice/ethanol bath for 15 min.

17. Collect the DNA by centrifugation (15 min, 12 000 g).

18. Wash the pellet with 80% ethanol.

19. Dissolve the pellet, after drying, in 50 μl TE buffer.

20. Check the amount and purity of the DNA by electrophoresis of a 1 μl aliquot on a 0.6% agarose minigel.

The DNA corresponding to inserts containing a Km^R fragment is isolated and ligated into an appropriately cleaved vector. The ligation reaction is subsequently used for transformation, and by plating a suitable number of double resistant clones (Amp^R, Km^R), a second library of mutated plasmids is established. These colonies are pooled, and the plasmid DNA is extracted. This mixture of plasmids, prepared as described in *Protocol 12*, is called DNA pool II. It consists exclusively of plasmids carrying mutations within the DNA of interest. Typical results are shown in *Figure 3*.

Protocol 12. Generation of DNA pool II

1. Pipette 1 μg of mutated inserts isolated from DNA pool I (*Protocol 11*, step 19) into a 1.5 ml microcentrifuge tube.

2. Add 1 μg of vector DNA cleaved with restriction enzymes A and B.

3. Add 5 μl 10 × ligase buffer (see *Protocol 7*), 3 μl 10 mM ATP and double distilled water to 49 μl.

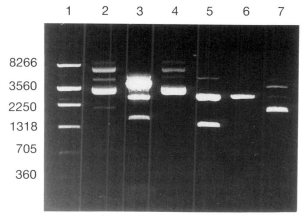

Figure 3. Intermediates from the construction of linker scanning mutations (II). Samples were taken at various stages of the linker scanning mutagenesis procedure and subsequently analysed on a 0.6% agarose gel. Lane 1, size standards; lane 2, supercoiled plasmid DNA extracted from pool I; lane 3, same DNA as shown in lane 2 after digestion with restriction enzymes A and B (the 1.7 kb band corresponds to inserts containing a Km^R fragment); lane 4, plasmid DNA extracted from pool II; lane 5, same DNA as shown in lane 4 after digestion with restriction enzyme X; lane 6, same DNA as shown in lane 5 after removal of the Km^R fragment; lane 7, same DNA as shown in lane 6 after circularization by T4 DNA ligase.

4. Start the reaction by adding 40 units T4 DNA ligase (New England Biolabs). The ligation is carried out overnight in a 15 °C water bath.

5. Use a 1 μl aliquot for an analytical colony transformation (*Protocol 9*). Select for ampicillin and kanamycin resistance.

6. Perform a preparative transformation with 10 μl of the ligation reaction (*Protocol 9* and Section 2.5).

7. Plate between 50 000 and 100 000 transformants on four large LB plates (Nunc Petri dishes, 24 × 24 cm) containing ampicillin and kanamycin each at 100 μg/ml.

8. Incubate the plates upside down for 20 to 24 h at 37 °C.

9. Wash the colonies off the plates using LB medium and a L-shaped glass. rod.

10. Collect the bacteria by centrifugation (5 min, 5000 g).

11. Extract the plasmid DNA ('DNA pool II') according to *Protocol 10*.

2.7 Removal of the kanamycin resistance fragment

At this stage the Km^R fragment is no longer required. In order to remove it, pool II DNA is digested with restriction enzyme X (see *Figure 1*) and the products are separated by gel electrophoresis. The band containing mutated 'starting plasmids' is cut out and the DNA is recovered from the gel (see *Protocol 13*). This DNA is subsequently circularized *in vitro* by T4 DNA ligase (*Protocol 14*). The reaction is performed on a preparative scale for reasons discussed in Section 3.5. We recommend performing first a small scale (0.5 ml) reaction before carrying out the ligation reaction of the entire sample. After a successful circularization reaction, the majority of the reaction products should consist of monomeric closed circles and monomeric nicked circles (see *Figure 3*, lanes 6 and 7). As both DNA species are suitable substrates for the following size fractionation step, the reaction products can be used directly without further purification. Typical results are shown in *Figure 3*.

Protocol 13. Removal of the kanamycin resistance (Km^R) fragment

1. Pipette 100 μg plasmid DNA (*Protocol 12*, step 11) into a 1.5 ml microcentrifuge tube.
2. Add 25 μl of the appropriate 10 × restriction enzyme buffer.
3. Add 200 units of the restriction enzyme X (see *Figure 1*) and double distilled water to 250 μl.
4. Incubate for 2 h at the appropriate temperature.
5. Check the digestion by electrophoresis of a 1 μl aliquot on a 0.6% agarose minigel.
6. Stop the digestion by adding EDTA to a final concentration of 10 mM.
7. Separate the two main reaction products (vector containing mutated insert from kanamycin resistance fragment, see *Figure 3*) on a 0.6% low melting point agarose gel (*Protocol 11*).
8. Excise the band corresponding to mutated 'starting plasmids'.
9. Extract the DNA from the low melting point agarose as described in *Protocol 11* and dissolve the DNA finally in 100 μl TE buffer.

Protocol 14. Large-scale intramolecular ligation (circularization)

1. Pipette 50 μg of 'pool II DNA' after the removal of the Km^R fragment (*Protocol 13*, step 9) into a 50-ml polypropylene tube.
2. Add 5 ml 10 × ligase buffer (see *Protocol 7*), 250 μl 100 mM ATP and double distilled water to give a volume of 44.75 ml.
3. Start the reaction by adding 2000 units (= 5 μl) T4 DNA ligase (New England Biolabs).
4. Incubate the tube overnight in a 15°C water bath.

5. Add 1 ml 0.5 M EDTA.

6. Reduce the volume to about 5 ml by several extractions with 2-butanol.

7. Remove residual traces of 2-butanol by two ether extractions.

8. Add 500 µl 3 M sodium acetate (pH 9.0) and 10 ml ethanol.

9. Place the tube for 30 min in a dry-ice/ethanol bath.

10. Collect the DNA by centrifugation (15 min, 12 000 g).

11. Resuspend the pellet in 300 µl TE buffer.

12. Add 150 µl 7.5 M ammonium acetate and 1000 µl ethanol.

13. Precipitate the DNA for 15 min in a dry-ice/ethanol bath.

14. Centrifuge for 15 min at 12 000 g.

15. Wash the pellet with 500 µl 80% ethanol.

16. Dissolve the DNA in 200 µl TE buffer.

17. On a 0.6% agarose gel separate aliquots of DNA before and after ligation to check the efficiency of this step (see *Figure 3*, lanes 6 and 7).

2.8 Size fractionation of mutated inserts

After the removal of the Km^R fragment, a stringent size fractionation of the mutated inserts is performed (*Protocol 15*). This step leads to a further enrichment of inserts containing correct LS mutations. The mutated inserts are excised from the *in vitro* circularized DNA with restriction enzymes A and B (see *Figure 1*) and the vector DNA is removed by agarose gel electrophoresis. The inserts are recovered from the agarose gel and subsequently separated on a non-denaturing polyacrylamide gel (see *Figure 4*). From the smear representing the pool of mutated inserts a narrow band, which contains DNA of about wild-type length, is excised.

Protocol 15. Size fractionation of mutated inserts

1. Digest about 50 µg of mutated 'starting plasmids' circularized *in vitro* (*Protocol 14*, step 16) with restriction enzymes A and B (see *Figure 1*) in order to excise the mutated inserts.

2. Separate the inserts from the vector DNA on a 1% agarose gel run in 1 × Tris-acetate buffer (*Protocol 11*).

3. Cut out the band representing the pool of mutated inserts.

4. Recover the DNA from the gel slices by isotachophoresis (*Protocol 16*).

5. Pour a 1 mm thick and 40 cm long non-denaturing 5% polyacrylamide gel.

6. Load about 1 µg mutated inserts in a 5 mm wide slot.

7. Load in adjacent lanes internal size standards (wild-type inserts excised

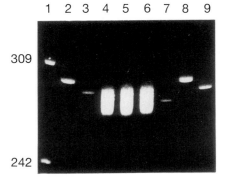

Figure 4. Intermediates from the construction of linker scanning mutations (III). For the performance of the size fractionation step, the mutated inserts were excised and separated on a 5% polyacrylamide gel. Lane 1, external size standards; lanes 2 and 8, internal size standard, wild-type inserts excised as a 289 bp fragment; lanes 3 and 9, internal size standard, wild-type inserts excised as a 281 bp fragment with restriction enzymes A and B; lanes 4 to 6, pool of mutated inserts excised with restriction enzymes A and B; lane 7, internal size standard, wild-type inserts excised as a 271 bp fragment.

Protocol 15. *continued*

with the same restriction enzymes as the mutated ones as well as with some other combination of restriction enzymes in order to obtain inserts that are a few base-pairs shorter or longer).

8. Run the gel in $0.5 \times$ TBE buffer (*Protocol 7*) at 750 V until the marker dyes have reached appropriate positions (Bromophenol Blue comigrates with a fragment of about 65 bp, xylene cyanol comigrates with a fragment of about 260 bp).

9. Stain the gel for 10 min in running buffer containing ethidium bromide (1 μg/ml) to visualize the DNA. Shake gently during staining.

10. Excise from the smear representing the pool of mutated inserts (see *Figure 4*) a narrow band containing mutated inserts of about wild-type length, using wild-type inserts run in parallel lanes as a reference.

11. Excise in addition some bands containing mutated inserts slightly smaller or bigger than wild-type inserts.

12. Recover the DNA from the gel slices by isotachophoresis (*Protocol 16*).

The removal of the vector DNA is a prerequisite for a successful size fractionation of the mutated inserts. Otherwise, one would overload the polyacrylamide gel and a considerable loss of resolution would result. The use of internal size standards is also required for an effective size fractionation. The reason is that small DNA fragments of exactly the same size, but differing in sequence, can migrate at different positions in a polyacrylamide gel.

The size fractionation procedure as described includes two gel steps. Since only limited amounts of DNA are available, an efficient procedure for the recovery of DNA is required. We recommend using isotachophoresis (see *Protocol 16*) since it allows quantitative recovery of DNA from agarose as well as polyacrylamide gels (28). A simple device for isotachophoresis is depicted schematically in *Figure 5*.

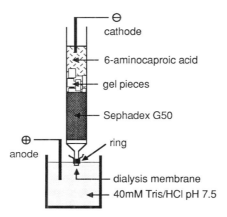

Figure 5. Recovery of DNA from agarose or polyacrylamide gel slices by isotachophoresis. The diagram shows a simple apparatus for isotachophoresis. For explanations, see Section 2.8 and *Protocol 16*.

Protocol 16. Isotachophoresis

1. Pack a small column (e.g. Bio-Rad Econo column of appropriate size) with autoclaved Sephadex G-50 fine (Pharmacia) in 40 mM Tris–HCl (pH 7.5).

2. Wash the column with at least 20 vol. of 40 mM Tris–HCl (pH 7.5).

3. Close the column outlet using a small piece of dialysis membrane and a small ring (cut off female Luer fitting). Avoid trapping air bubbles in the outlet.

4. Fix the column so that the outlet dips into a 100 ml glass beaker filled with 40 mM Tris–HCl (pH 7.5).

5. Remove the residual buffer in the column.

6. Place the DNA-containing gel pieces (a large piece of gel should first be cut into small pieces with a scalpel) on top of the Sephadex bed. The total volume of gel pieces should be smaller than the volume of the Sephadex.

7. Add 25 µl of a 0.1% phenol red solution.

8. Fill the column to the top with a 100 mM solution of 6-aminocaproic acid in double distilled water.

Protocol 16. *continued*

9. Connect a pair of electrodes to a power supply. Place the cathode into the upper reservoir of the column filled with 6-aminocaproic acid and the anode into the beaker filled with Tris buffer.

10. Apply between 100 and 500 V (1 to 10 mA) depending on the size of the column. During isotachophoresis the phenol red and the electroeluted DNA form a sharp band slowly migrating towards the anode.

11. Switch the power supply off as soon as the phenol red is 0.5 to 1 cm away from the lower end of the Sephadex bed.

12. Remove the 6-aminocaproic acid from the upper buffer reservoir.

13. Substitute with 40 mM Tris–HCl (pH 7.5).

14. Remove the dialysis membrane.

15. Immediately collect fractions consisting of 1 to 4 drops.

16. Stop collecting when the phenol red has disappeared from the column. The DNA is in those fractions preceding the phenol red.

17. Identify the DNA containing fractions either by measuring the radiation if the DNA was radiolabelled or by a spot-test. For the spot-test, pipette an appropriate number of 2 μl aliquots of TE buffer containing ethidium bromide (2μg/ml) onto a piece of Saran Wrap placed on the screen of an UV transilluminator. Remove 2 μl aliquots from the collected fractions and mix them with the droplets on the transilluminator. DNA-containing fractions show an increased fluorescence upon UV illumination which is correlated with the DNA concentration in the samples.

18. Pool the DNA-containing fractions and recover the DNA by ethanol precipitation. Add a carrier or concentrate the sample by several extractions with 2-butanol if the DNA concentration is too low.

The size fractionated insert DNA, is ligated into an appropriately cleaved vector and used to transform bacteria (*Protocol 17*). At this stage, at least 1 in 10 clones should contain a plasmid with a correct LS mutation.

Protocol 17. Ligation of size fractionated inserts into a vector and subsequent transformation

1. Pipette 500 ng appropriately cleaved vector DNA into a 1.5 ml microcentrifuge tube.

2. Add 50 to 100 ng mutated inserts of about wild-type length (*Protocol 15*, step 12).

3. Add 2 μl 10 × ligase buffer (*Protocol 7*), 1 μl of a 10 mM ATP solution and double distilled water to 19 μl.

4. Start the reaction by adding 40 units T4 DNA ligase (New England Biolabs).

5. Incubate the tube overnight at 15°C.

6. Use a 1 μl aliquot to perform a colony transformation (*Protocol 9*). If a pUC type vector and a *lacZ⁻* strain such as JM109 have been used, colonies harbouring a recombinant plasmid can easily be identified by their white colour on LB plates containing ampicillin (100 μg/ml), IPTG (2 ml of a 100 mM solution in water per litre) and X-gal (3 ml of a 2% solution in dimethylformamide per litre).

7. Pick white colonies and use them to inoculate 'miniprep' cultures containing ampicillin (100 μg/ml) and extract the plasmid DNA as described in *Protocol 18*.

2.9 Identification of putative linker scanning mutations on the basis of the topoisomer pattern

Of the clones obtained after the size fractionation step, approximately 1 in 10 contain a plasmid with a correct LS mutation. These clones are identified in the following step by a simple enzymatic screening procedure. First, miniprep plasmid DNA is isolated from individual clones (*Protocol 18*). These mutated plasmids and the 'starting plasmid' (wild-type) are subsequently relaxed to completion by topoisomerase I (*Protocol 19*) and the resulting topoisomers are separated on agarose gels (*Protocol 20*). This procedure allows the distinction of plasmids differing in length by a single base-pair (29). It exploits the fact, that topoisomers can be resolved into discrete bands by gel electrophoresis. Insertion or deletion of x base-pairs ($x = 1, 2, 3, \ldots$) leads to a clearly visible shift of these bands (see *Figure 6*; for a detailed discussion of the theoretical background see ref. 29). Plasmids with a correct LS mutation are therefore expected to show the wild-type topoisomer pattern. In mutated plasmids displaying the wild-type topoisomer pattern, the position of the linker is approximately determined by restriction mapping. Mutants carrying the linker at a desired position are finally sequenced using a rapid dideoxy sequencing procedure for supercoiled plasmid DNA (30, 31). The topoisomerase I screening procedure guarantees that the majority of clones chosen for DNA sequencing are correct LS mutants without insertions or deletions. *Figure 7* shows a representative series of linker scanning mutations which was obtained by the procedure described in this chapter.

Protocol 18. Small-scale preparation of plasmid DNA ('miniprep')

1. Inoculate 2 ml LB medium (see *Protocol 8*) containing the appropriate antibiotic (e.g. ampicillin 100 μg/ml) with a single bacterial colony.

2. Shake vigorously overnight at 37°C.

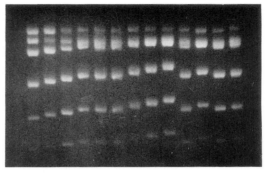

Figure 6. Effect of small deletions or insertions on the topoisomer pattern of a plasmid. Supercoiled plasmid DNAs from the wild-type and from selected mutants were completely relaxed with topoisomerase I and separated on an agarose gel as described in *Protocols 19* and *20*. The lanes contain wild-type DNA (wt), DNA from a correct LS mutant of exactly the same size as the wild-type (LS) as well as DNAs from incorrect LS mutants, which show in comparison to the wild-type either deletions (LS − 1, LS − 2,...) or insertions (LS + 1, LS + 2,...) of a few base-pairs.

Protocol 18. *continued*

3. Transfer 1.5 ml to a microcentrifuge tube. Pipette the remainder of the suspension into another microcentrifuge tube and store it at 4°C. After having identified a LS mutant the stored suspension can be used to inoculate a large-scale liquid culture of this particular clone.

4. Pellet the bacteria for 1 min at 12 000 g.

5. Aspirate the supernatant.

6. Resuspend the bacteria thoroughly in 400 μl STET buffer (*Protocol 10*) by pipetting.

7. Add 32 μl of a freshly prepared lysozyme solution (10 mg/ml in H_2O).

8. Mix and place the tube immediately for 50 sec in a boiling-water bath.

9. Place the tube on ice for 1 to 2 min.

10. Centrifuge for 15 min at 12 000 g.

11. With a pipette remove the bacterial debris which should form a large soft pellet.

12. Add 400 μl isopropanol to the supernatant, mix.

13. Precipitate the nucleic acids for 1 h at −20°C.

14. Centrifuge for 15 min at 12 000 g.

15. Aspirate the supernatant.

16. Wash the pellet with 500 μl 80% ethanol.

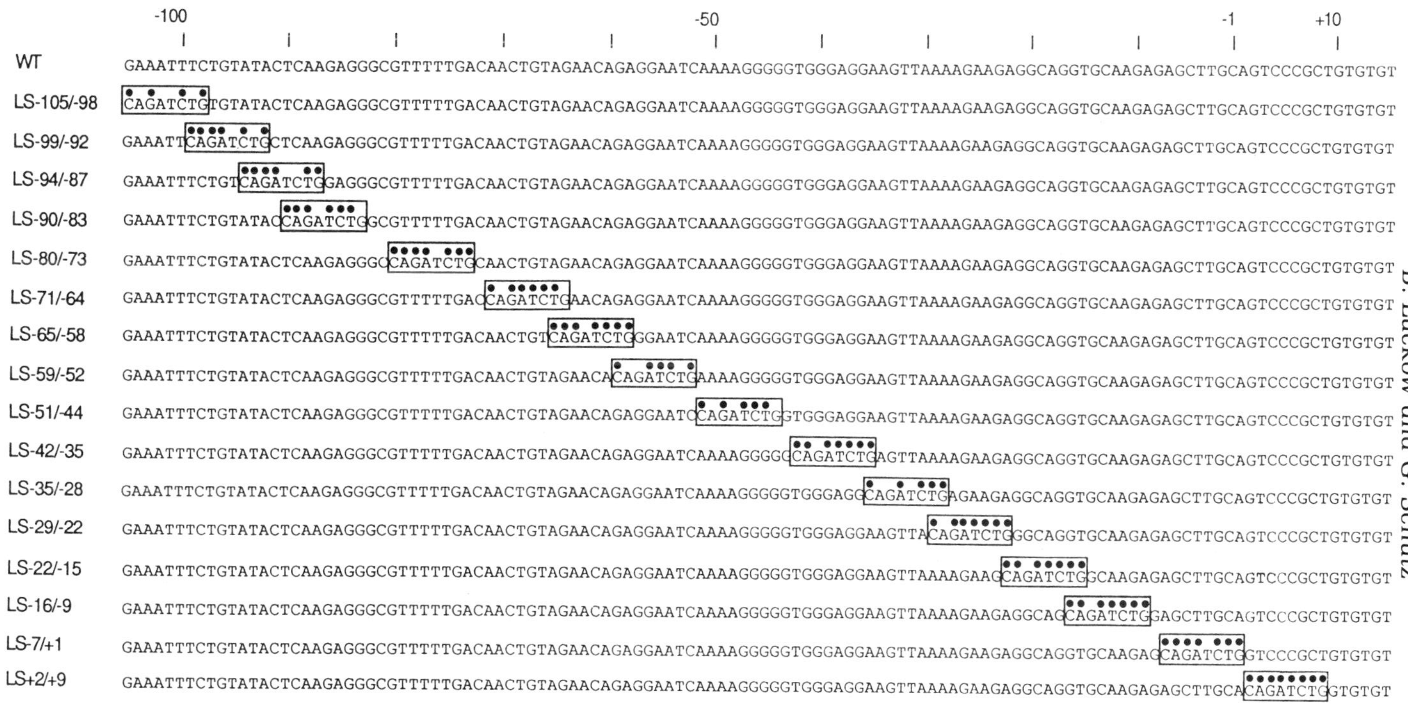

Figure 7. Linker scanning mutations in the promoter of the chicken lysozyme gene. The sequences of 16 different LS mutations and the wild-type sequence from position −105 to +15 are shown. All LS mutations depicted in this figure have been constructed with the help of the procedure described in this chapter. The sequence of the *Bgl*II linker, which was used for linker scanning mutagenesis, is boxed. Mutated nucleotides are indicated by a dot. The mutations are designated according to the pattern LSn/m, where n and m specify the position of the first and the last nucleotide of the wild-type sequence that has been replaced by the linker sequence.

125

WT
LS-105/-98
LS-99/-92
LS-94/-87
LS-90/-83
LS-80/-73
LS-71/-64
LS-65/-58
LS-59/-52
LS-51/-44
LS-42/-35
LS-35/-28
LS-29/-22
LS-22/-15
LS-16/-9
LS-7/+1
LS+2/+9

−100   −50   −1   +10

Protocol 18. *continued*

17. Centrifuge for 3 min at 12 000 *g*.

18. Aspirate the supernatant carefully.

19. Dry the pellet for a few minutes in the air.

20. Dissolve the pellet in 50 μl TE buffer.

Protocol 19. Relaxation of supercoiled plasmid DNA by topoisomerase I

1. Prepare a pre-mix containing per reaction 1 μl 10 × topoisomerase I buffer (100 mM Tris–HCl (pH 8.0), 2.5 M NaCl, 1 mM EDTA), 6 μl double distilled water and 1 μl of an appropriate topoisomerase I dilution. The amount of enzyme needed to relax completely about 300 ng of 'miniprep' plasmid DNA is determined in a pilot experiment.

2. Pipette 8 μl of this pre-mix into a 1.5 ml microcentrifuge tube.

3. Add 2 μl (≈300 ng) 'miniprep' plasmid DNA (*Protocol 18*, step 20), mix by vortexing, centrifuge for a few seconds.

4. Incubate for 2 h at 37°C.

5. Stop the reaction by adding 10 μl TE buffer containing 0.1% SDS. After the addition of loading buffer with xylene cyanol as tracking dye, the sample can be analysed immediately on an appropriate gel.

Protocol 20. Topoisomer pattern analysis of relaxed plasmids

1. Combine 4.5 g agarose, 7.5 ml 40 × running buffer (1.6 M Tris–HCl (pH 7.9), 200 mM sodium acetate, 40 mM EDTA) and double distilled water in a 500 ml Erlenmeyer flask. The total volume should be 300 ml.

2. Heat the flask in a microwave oven until the agarose has completely dissolved.

3. Replace the evaporated water and cool the flask down in a 50°C water bath.

4. Pour a 20 × 24 × 0.6 cm horizontal gel and use a 2-mm thick comb with 25 teeth each 4 mm wide if available.

5. Remove the comb after the gel has set.

6. Fill the buffer reservoirs with freshly prepared 1 × running buffer.

7. Load relaxed plasmid DNAs (*Protocol 19*, step 5). We recommend loading 20 mutant DNAs and 5 wild-type DNAs on each gel.

8. Run the gel at 100 V/100 mA for about 24 h at room temperature, until the xylene cyanol has migrated 20 cm. The buffer must be circulated with a pump during this time to avoid the generation of a pH gradient.

9. Stain the DNA by immersing the gel for 30 min in water containing ethidium bromide (1 µg/ml). Shake gently during staining.

10. Photograph the gel for a careful analysis of the different topoisomer patterns.

Protocol 21. Sequence determination of supercoiled plasmid DNA

A. **Template denaturation**

1. Pipette 16 µl 'miniprep' DNA (1–2 µg, ≈1 pmol) prepared according to *Protocol 18* into a 1.5 ml microcentrifuge tube.

2. Add 4 µl denaturation solution (1 M NaOH, 1 mM EDTA).

3. Vortex and incubate for 5 min at 65°C.

4. Neutralize by adding 2 µl of a 2 M ammonium acetate solution adjusted to pH 4.5 with acetic acid.

5. Precipitate the DNA with 60 µl ethanol for 10 min in a dry-ice/ethanol bath.

6. Centrifuge for 10 min at 12 000 g and 4°C.

7. Wash the pellet with 500 µl ice-cold 80% ethanol.

8. Dry the pellet in a SpeedVac for 1 to 2 min and use the DNA immediately for sequencing.

B. **Primer annealing**

9. Pipette 6 µl double distilled water, 2 µl 5 × sequencing buffer[a] and 2 µl (= 1 pmol) of the appropriate primer (e.g. M13 universal sequencing primer) into the tube with the denatured plasmid DNA.

10. Dissolve the DNA by vortexing, then centrifuge for a few seconds.

11. Incubate for 15 min at 37°C.

C. **Labelling reaction**

12. Add 2 µl labelling mix (1.5 µM dCTP, 1.5 µM dGTP, 1.5 µM dTTP), 1 µl of a 100 mM DTT solution, 0.5 µl [α-^{35}S]dATP (>1000 Ci/mmol, 8 µCi/µl) and 2 µl (=2 units) diluted T7 DNA polymerase.

13. Mix and incubate for 5 min at room temperature.

D. **Extension-termination reaction**

14. Dispense 3 µl aliquots of the labelling reaction (step 13) to 4 wells of a

Protocol 21. *continued*

microsample plate (e.g. Pharmacia No. 2010–700) labelled G, A, T, and C.

15. Add 2.5 μl of the appropriate termination mix[b] to each well (i.e. G termination mix to well G, A termination mix to well A, etc.). Mix by pipetting.

16. Incubate the microsample plate for 10 min at 37 °C.

17. Stop the reactions by adding 5 μl formamide loading buffer (95% deionized formamide, 20 mM EDTA, 0.1% xylene cyanol, 0.1% Bromophenol Blue) to each well. Mix by pipetting.

18. Heat the samples to 80 °C for 5 min immediately before loading them on a standard sequencing gel. Load 2–3 μl per lane.

[a] 5× sequencing buffer is 200 mM Tris–HCl (pH 7.5), 100 mM MgCl$_2$, 250 mM NaCl.
[b] A termination mix 80 μM dATP, 80 μM dCTP, 80 μM dGTP, 80 μM dTTP, 8 μM ddATP, 50 mM NaCl; C termination mix, G termination mix, T termination mix. Same composition as A termination mix, but the ddATP is replaced by ddCTP, ddGTP or ddTTP, respectively.

The following paragraphs contain further comments on the procedures mentioned above.

(i) The plasmid minipreps should be performed according to the protocol developed by Holmes and Quigley (25). This protocol involves a minimal number of manipulations and the plasmid DNA obtained by this procedure is a good substrate for restriction enzyme digestions, topoisomerase I and sequencing reactions.

(ii) The relaxation of plasmids should be performed in a warm room rather than using a heating block to avoid significant changes in the ionic conditions due to condensation of water on the lid of the tube during the incubation. If a sample of plasmid DNA is not fully relaxed after 2 h, the amount of enzyme has to be increased rather than extending the incubation time, because topoisomerase I, at least in our hands, is active for no more than 2 h. We have always used topoisomerase I prepared from calf thymus by Dr H.-P. Vosberg, Heidelberg, but topoisomerase I from other sources might work as well.

(iii) Concerning the analysis of topoisomers by gel electrophoresis, several aspects have to be considered.

• The topoisomer bands will become fuzzy if the gel is run at a higher voltage than 100 V. The same is true if the running buffer is reused.

• Topoisomer gels should not be substantially thicker than 6 mm, otherwise the staining of the DNA becomes difficult.

• The staining period should be restricted to 30 min, otherwise an increased background fluorescence will be obtained.

- It may be possible to substitute Tris-acetate with Tris-borate buffer, which does not require pumping, however, we have not checked this alternative.

(iv) Sequencing reactions using 'miniprep' DNA are frequently of the same quality as those using DNA purified by two CsCl/ethidium bromide gradients. However, lower quality sequencing reactions are usually sufficient to determine the exact genotype of a LS mutant because, due to the experimental strategy, only the linker and a few surrounding base-pairs need to be readable unambiguously. The use of microsample plates simplifies and accelerates the work considerably, especially if several plasmids are sequenced in parallel.

3. General aspects of linker scanning mutagenesis

3.1 Selecting a suitable DNA fragment

In principle, any piece of double-stranded DNA can be used as a substrate for linker scanning mutagenesis. However, there are certain limitations with respect to the size of the fragment which is to be mutated; it should have a minimum length of about 50 bp and a maximum length of about 300 bp. The reasons for these size limits are twofold. First, it may not be worthwhile analysing a DNA fragment by linker scanning mutagenesis if it is considerably shorter than 50 bp, in which case single point mutations may be more helpful. Second, if a fragment is longer than 300 bp, then it becomes difficult to perform an effective size fractionation. The success of this step is essential for the topoisomerase I screening procedure since it reduces the number of clones to be screened, and excludes false-positive clones, which have deletions of, for example, 10 bp (see *Figure 5*).

3.2 Selecting a suitable vector

Before one can begin linker scanning mutagenesis the DNA of interest has to be inserted into a suitable vector thereby generating the so-called 'starting plasmid'. The vector has to meet the following requirements. First, it should be relatively small in size because the resolution of different topoisomers into discrete bands is faster and better with smaller plasmids and the percentage of mutations within the DNA of interest is increased if a smaller vector is used. Second, the vector should contain a polylinker region. This allows the DNA of interest to be inserted so that several unique restriction sites are available at each end of the fragment facilitating the generation of internal size standards for the size fractionation step. Plasmids such as pUC18 and pUC19 (32) are well-suited as vectors for linker scanning mutagenesis;

- they give high yields of plasmid DNA;

- recombinant plasmids can be identified by a simple colour screening procedure;
- the universal M13 sequencing primers can be used for sequence determination.

3.3 Selecting a suitable linker

The linker used for linker scanning mutagenesis should be chosen so that the recognition sequence it carries is not present within the 'starting plasmid'. The length of the linker determines the degree of resolution which can be achieved. In addition, it determines the minimal number of clones needed to saturate the DNA of interest. Linkers containing recognition sequences of commonly used restriction enzymes are commercially available with a length of either 8, 10, or 12 bp. The efficiency of linker ligation appears to be somewhat higher with dodecameric linkers than with corresponding octameric linkers.

3.4 Essential steps

Two steps appear to be essential for the mutagenesis procedure described in this chapter. First, the initial linearization of the 'starting plasmid' has to be as random as possible. This is achieved by treating the plasmids with formic acid, exonuclease III, and nuclease S1. As every base-pair contains a purine, every base-pair represents a potential target for acid-catalysed apurination, thus ensuring that cuts are evenly dispersed throughout the DNA of interest. The second important feature is the possibility of identifying correct LS mutants by a fast and simple enzymatic screening procedure. This step considerably reduces the number of clones which have to be sequenced to establish a series of LS mutations. For example, in one study, of 26 mutants displaying the wild-type topoisomer pattern, 21 proved to be correct LS mutants (17). However, a size fractionation step has to be performed in advance to minimize the number of clones that have to be screened, as well as to exclude false-positive clones with deletions of, for example 10 bp, which display a topoisomer pattern identical to wild-type (see *Figure 5*).

3.5 Potential problems

Impure 'miniprep' DNA requires about 30 times more topoisomerase I for complete relaxation than DNA purified by a caesium chloride/ethidium bromide gradient. For this reason large amounts of topoisomerase I are required. The enzyme is commercially available but is expensive; so if cost is a problem, one should consider preparing topoisomerase I. Since there is no need for highly purified enzyme, it might be sufficient to use an abbreviated version of a standard topoisomerase I purification protocol (33). Another alternative is to use chloroquine to relax supercoiled plasmid DNA (34).

Although this is effective, topoisomers generated by chloroquine give less satisfactory gel resolution compared with topoisomers produced by topoisomerase I. A 1.5% agarose gel containing chloroquine at a concentration of 100 μg/ml allows the identification of LS mutants differing in length from the wild-type by two or more base pairs.

A more serious concern is the genetic stability of pooled populations of bacterial transformants. We have noticed that the distribution of linker insertion becomes more non-random with additional amplifications of the original mutant library (pool I). Mutants having the linker inserted at some specific sites within the DNA of interest have a selective advantage and are enriched after each transformation step. The problem could be minimized by reducing the number of transformations involving a pool of mutant plasmids. For this reason, our strategy for LS mutagenesis contains a large-scale circularization reaction after the removal of the Km^R fragment from pool II DNA. Using aliquots of the same ligation reaction for the transformation of two different *E. coli* strains we have made the following, unexpected, observation. Strain JM109 showed a random distribution of linker insertion when 24 independent transformants were analysed, whereas a strong 'hot spot' for linker insertion was evident with DH5. Keeping the distribution of linkers within the DNA of interest as random as possible is obviously crucial for our mutagenesis strategy, otherwise the number of clones that must be screened for a complete series of LS mutations is increased considerably.

3.6 Advantages

The procedure for linker scanning mutagenesis described in this chapter offers several clear advantages.

- The number of mutations which have to be sequenced in order to find a correct LS mutation has been reduced considerably in comparison to other methods. In addition, there is no need to construct each LS mutation separately by isolating and ligating appropriate 5'- and 3'-deletion fragments. Both features facilitate and accelerate the construction of LS mutations. The procedure is therefore especially suited for systematic studies requiring many different LS mutations.

- Although designed primarily for the construction of correct LS mutations, the method can also be used to isolate mutations causing insertions or deletions of a defined number of base-pairs. Such mutations might be helpful in analysing the stereospecific alignment of regulatory sequences.

- The method involves only simple manipulations and does not require expensive equipment such as a DNA synthesizer.

3.7 Time requirements

The construction of a representative series of LS mutations is a time-consuming task. Assuming that the work is done by a single person, one month should be sufficient to perform all the steps from the apurination of the 'starting plasmid' to the size fractionation of the mutated inserts. The topoisomerase I screening procedure requires additional time mainly determined by the number of clones which must be screened for a complete series of LS mutations. This number depends on:

- the size of the DNA of interest;
- the size of the linker;
- the randomness of the linker distribution;
- the efficiency of the size fractionation step.

In any case it will be of the order of a few hundred. The screening of 1000 clones by topoisomerase I requires 1–2 months for a single person; this period includes the final characterization of LS mutations by DNA sequencing. The topoisomerase I screening procedure, although very simple, is therefore the most time-consuming step of the entire strategy.

Acknowledgements

The authors thank Dr H. P. Vosberg for considerable amounts of topoisomerase I, Drs G. Kelsey, M. Nichols and A. F. Stewart for their comments on the manuscript, W. Fleischer for photography, and C. Schneider as well as P. DiNoi for typing.

References

1. McKnight, S. L. and Kingsbury, R. (1982). *Science*, **217**, 316.
2. Charnay, P., Mellon, P., and Maniatis, T. (1985). *Molecular and Cellular Biology*, **5**, 1498.
3. Zajchowski, D. A., Boeuf, G., and Thimmappaya, B. (1985). *EMBO Journal*, **4**, 1293.
4. Murthy, S. C., Bhat, G., and Thimmappaya, B. (1985). *Proceedings of the National Academy of Sciences of the USA*, **82**, 2230.
5. Haltiner, M., Smale, S. T. and Tjian, R. (1986). *Molecular and Cellular Biology*, **6**, 227.
6. Buetti, E. and Kühnel, B. (1986). *Journal of Molecular Biology*, **190**, 367.
7. Windle, J. J. and Sollner-Webb, B. (1986). *Molecular and Cellular Biology*, **6**, 4585.
8. Addison, W. R. and Kurtz, D. T. (1986). *Molecular and Cellular Biology*, **6**, 2334.
9. Bhat, G., SivaRaman, L., Murthy, S., Domer, P., and Thimmappaya, B. (1987). EMBO *Journal*, **6**, 2045.

10. Jones, M. H., Learned, R. M., and Tjian, R. (1988). *Proceedings of the National Academy of Sciences of the USA*, **85**, 669.
11. Ayer, D. E. and Dynan, W. S. (1988). *Molecular and Cellular Biology*, **8**, 2021.
12. Sharp, S. J. and Garcia, A. D. (1988). *Molecular and Cellular Biology*, **8**, 1266.
13. Railey, J. F. and Wu, G. J. (1988). *Molecular and Cellular Biology*, **8**, 1147.
14. Crowe, D. T. and Tsai, M. J. (1989). *Molecular and Cellular Biology*, **9**, 1784.
15. Luckow, B. and Schütz, G. (1989). *Nucleic Acids Research*, **17**, 8451.
16. Haltiner, M., Kempe, T., and Tjian, R. (1985). *Nucleic Acids Research*, **13**, 1015.
17. Luckow, B., Renkawitz, R., and Schütz, G. (1987). *Nucleic Acids Research*, **15**, 417.
18. Lichtler, A. and Hager, G. L. (1987). *Gene Analysis Techniques*, **4**, 111.
19. Zoller, M. J. and Smith, M. (1983). In *Methods in Enzymology* (ed. R. Wu, L. Grossman, and K. Moldave), Vol. 100, p. 468. Academic Press, New York.
20. Barany, F. (1985). *Proceedings of the National Academy of Sciences of the USA*, **82**, 4202.
21. Maxam, A. M. and Gilbert, W. (1980). In *Methods in Enzymology* (ed. L. Grossman and K. Moldave), Vol. 65, p. 499. Academic Press, New York.
22. Rogers, S. G. and Weiss, B. (1980). In *Methods in Enzymology* (ed. L. Grossman and K. Moldave), Vol. 65, p. 201. Academic Press, New York.
23. Wiegand, R. C., Godson, G. N., and Radding, C. (1975). *Journal of Biological Chemistry*, **250**, 8848.
24. Oka, A., Sugisaki, H., and Takanami, M. (1980). *Journal of Molecular Biology*, **147**, 217.
25. Holmes, D. S. and Quigley, M. (1981). *Analytical Biochemistry*, **114**, 193.
26. Hanahan, D. (1985). In *DNA Cloning – A Practical Approach* (ed. D. M. Glover), Vol. I, p. 109. IRL Press, Oxford and Washington DC.
27. Dower, W. J., Miller, J. F., and Ragsdale, C. W. (1988). *Nucleic Acids Research*, **16**, 6127.
28. Öfverstedt, L-G., Hammarström, K., Balgobin, N., Hjerten, S., Pettersson, U., and Chattopadhyaya, J. (1984). *Biochimica et Biophysica Acta*, **782**, 120.
29. Wang, J. C. (1979). *Proceedings of the National Academy of Sciences of the USA*, **76**, 200.
30. Chen, E. Y. and Seeburg, P. H. (1985). *DNA*, **4**, 165.
31. Tabor, S. and Richardson, C. C. (1987). *Proceedings of the National Academy of Sciences of the USA*, **84**, 4767.
32. Yanisch-Perron, C., Vieira, J., and Messing, J. (1985). *Gene*, **33**, 103.
33. Schmitt, B., Buhre, U., and Vosberg, H. P. (1984). *European Journal of Biochemistry*, **144**, 127.
34. Shure, M., Pulleyblank, D. E., and Vinograd, J. (1977). *Nucleic Acids Research*, **4**, 1183.

7

Random chemical mutagenesis and the non-selective isolation of mutated DNA sequences *in vitro*

CATHERINE WALTON, R. KIM BOOTH, and PETER G. STOCKLEY

1. Introduction

This chapter illustrates examples of the types of problems to which random *in vitro* chemical mutagenesis lends itself. It outlines the most frequently used techniques of chemical mutagenesis, and describes in full experimental detail the two chemical treatments which are the simplest to apply and which generate the most nearly random distribution of mutations. The use of denaturing gradient gel electrophoresis to separate nucleic acids differing by as little as a single base-pair is also described in detail. Routine molecular biology techniques used in the construction of clones of DNA fragments and in the generation of single-stranded DNA for mutagenesis and sequencing are adequately documented elsewhere (1).

2. Applications of random chemical mutagenesis

2.1 Random versus directed mutagenesis *in vitro*

The generation *in vitro* of mutations at specified sites within a polynucleotide of known sequence is now routine. Particularly when used in combination with the increasingly rapid structural techniques available today, site-specific mutagenesis provides a very powerful tool for the manipulation of sequences and structures. Unfortunately, the productive use of these directed tools is often limited by the need to know which sites within a molecule should be investigated in detail; lack of sufficient structural information frequently means that a site-directed approach is inappropriate. Random mutagenesis, however, may be used to identify and investigate functionally interesting sites within a protein or nucleic acid sequence.

We have been studying the proteins which comprise the capsids of a family of related viruses. Individual molecules of these proteins perform several

135

independent biological functions. Protein molecules interact specifically with (a) other identical protein molecules within the viral coat, (b) a dissimilar protein molecule within the viral coat, (c) the chromosome of the virus within the viral particle and (d) the viral chromosome in a regulatory manner during development. No detailed structural information about these proteins, other than their primary sequences, is yet available. In an attempt to identify which regions of the primary amino acid sequences of these proteins are involved in each of their separate functions we have applied the approach of random saturation mutagenesis *in vitro*.

Even when the detailed three-dimensional structure of a protein is available, it is often impossible to deduce the degree to which particular amino acid residues play crucial roles either in determining functional aspects of a protein, or in maintaining its structure. Random mutagenesis offers the opportunity to screen proteins, particularly proteins whose detailed structure has previously been determined by crystallography or by magnetic resonance techniques, for structurally important residues or sites, without the selective bias which a directed approach necessarily introduces.

A third field of inquiry which lends itself to the application of random mutagenesis is the study of the nature of the regulatory and structural features of nucleic acid sequences, such as promoters, enhancers, ribosome binding sequences, transcription termination sequences, the structures and functions of transfer and ribosomal RNA, the processes involved in viral assembly, in primary transcript splicing and the initiation of nucleic acid replication. With problems of this type there is often little or no basis for making the specific predictions necessary for a site-directed approach. However, by using random mutagenesis followed by measurements of diminished, enhanced, or unaltered biological activity, a great deal of information can be obtained which will eventually suggest targets towards which a site-specific or random oligonucleotide approach can subsequently be directed.

3. Chemistry of random mutagenesis *in vitro*

3.1 Background to chemical mutagenesis *in vitro*

Over the past forty years, a large number of substances have been shown to react with nucleic acids, both *in vitro* and *in vivo*, and in so doing induce changes in the sequence of DNA that are subsequently replicated *in vivo*. However, the majority of these compounds cause various types of damage (2), and consequently only five of the best-understood treatments are currently widely employed in generating randomly distributed single-point mutations *in vitro*. These are nitrous acid, sulphurous acid (in the form of a bisulphite salt), hydroxylamine, hydrazine, and mild acid hydrolysis.

In practice, aqueous hydrazine and acid hydrolysis can produce all possible base substitutions in a highly reproducible and easily controlled manner (3,

4). For this reason, only the experimental procedures for these two treatments are described in detail (*Protocols 2* and *3*). Brief descriptions of nitrous acid, bisulphite, and hydroxylamine mutagenesis are included in this section. Their practical application is essentially the same as that described for hydrazine and mild acid treatments in *Protocols 2* and *3*.

3.2 Nitrous acid mutagenesis

As expected, the reaction between nitrous acid and the amines of purine and pyrimidine initially leads to the formation of their corresponding diazonium derivatives. These are unstable in aqueous solution, and rapidly decompose into their ketolic/enolic hydrolysis products. Thus guanine is converted into xanthine, adenine into hypoxanthine (inosine), and cytosine into uracil. The rates of these deamination reactions vary (in the approximate ratios of 1:2:6; guanine:adenine:cytosine) and increase rapidly with decreasing pH. On replication of nitrous acid treated DNA, cytidine is incorporated base-paired to hypoxanthine, thymidine base-paired to xanthosine, and adenosine base-paired to uridine. Nitrous acid therefore generates mutations of the transition (purine to purine and pyrimidine to pyrimidine) type (5).

3.3 Hydroxylamine mutagenesis

Under strongly acidic conditions (<pH 2.5), hydroxylamine acts as a powerful hydroxyl donor, the overall result of which is to deaminate cytosine to uracil. The naturally occurring derivatives of cytosine, 5-hydroxymethylcytosine and 5-[glycosyl]-hydroxymethylcytosine are also deaminated by hydroxylamine, although 5-methylcytosine is not. Under milder conditions (pH 6), an intermediate in the deamination of cytosine by hydroxylamine, hydroxyaminocytosine, is formed. This is stable and can form a non-Watson–Crick base-pair with adenine. Both mechanisms of hydroxylamine induced mutation give rise to C-G to A-T transition mutations (6).

3.4 Bisulphite mutagenesis

The reaction of bisulphite with cytosine is approximately a thousandfold faster than the deamination reactions between bisulphite and purines and is therefore the only one of interest here. The treatment of DNA with bisulphite yields the same sub-set of transition point mutations as treatment with hydroxylamine (7).

3.5 Hydrazine mutagenesis

Anhydrous hydrazine and aqueous solutions of hydrazine hydrate react specifically with the pyrimidine bases of DNA, leading to cleavage of the heterocyclic ring system (3). Subsequent spontaneous hydrolysis results in depurination of the nucleic acid at the modified base. The actual product left in the polynucleotide chain following treatment of DNA with hydrazine is not deoxyribose, but its 1-hydrazone derivative. If necessary, deoxyribose can be regenerated by treatment of the nucleic acid with benzaldehyde, although this is not usually necessary for the simple formation of point mutations from depyrimidinated DNA. Cytosine and thymidine are attacked at differing rates, in the ratio 1:3, respectively. The reaction is affected by ionic strength; the higher the cation concentration the faster the relative rate of attack of cytosine.

The detailed conditions used in the generation of mutations by hydrazine depyrimidination *in vitro* are described under *Protocol 3*.

3.6 Acid depurination of DNA

The N7-glycosidic bond between the purine bases and the C1 of the deoxyribose ring is far more labile to mild acid hydrolysis than any of the other bonds in DNA. By treatment with mild acid it is therefore possible to specifically excise purines from DNA (4). In practice, formic acid is generally used because of the ease with which all traces of the acid can subsequently be removed. A detailed description of the optimal conditions for the use of mild acid hydrolysis in the generation of mutations *in vitro* is given under *Protocol 2*.

4. Practical considerations for random chemical mutagenesis

4.1 Vector damage

To screen rapidly for mutations by DNA sequencing alone, a significant proportion of the molecules sequenced must be expected to be mutated. However, mutagenesis of an isolated cloned fragment of DNA exposes not only the fragment of interest to mutagenesis, but also essential vector sequences such as origins of replication and selective markers, making direct transformation of the mutated molecule impossible after very high doses of mutagen.

Early attempts to overcome this problem concentrated on the excision of the fragment to be altered from its vector, prior to treatment with the mutagen. Unfortunately, isolated fragments of double-stranded DNA, such as those generated by digestion with restriction endonucleases, do not easily lend themselves to mutagenesis *in vitro*. In general, those mutagens which have proved most suitable react much more readily with single-stranded than with double-stranded DNA (see Section 4.2). As a result, the ends of restriction fragments are attacked at rates typically hundreds of times faster than other parts of the molecule. In particular, extensive damage to the ends of duplex DNA makes subsequent recloning, for example, into a suitable sequencing vector, at best difficult and often impossible. The solution is to excise and isolate the sequence of interest after mutagenic treatment of the whole recombinant clone.

4.2 Distribution of induced mutations

Chemical mutagenesis of duplex DNA *in vitro* is highly non-random, almost certainly as a result of differential 'breathing' of the duplex exposing certain regions as transiently single-stranded more frequently than others. This leads to the distribution of mutations created by chemical treatments *in vitro* being biased by the sequence of the duplex DNA target.

This problem is considerably reduced by the use of single-stranded DNA as the mutagenesis substrate. However, it should be borne in mind that, even using a single-stranded substrate, the distribution of mutations is still not truly random. The lability of particular sites throughout the molecule and between particular treatments (8, 13) differs for reasons not fully understood. Most probably this arises from a combination of 'next neighbour' effects, and from the formation of regions of imperfectly base-paired duplex between different regions of the single-stranded molecule. Of particular interest is the observation that hydrazine depyrimidination of single-stranded DNA, preferentially gives rise to pyrimidine to pyrimidine transition mutations (ref. 8 and *Figure 1*). In practice, mutagenesis of both complementary strands of a sequence of interest overcomes such limitations (8).

A further bias in the distribution of mutants occurs on enrichment of mutagen-treated DNA for altered sequences (see Section 5). The enrichment procedure separates DNA sequences on the basis of differences in their melting temperatures from that of the wild-type sequence. Consequently, the products of the enrichment procedure are fractions of treated DNA enriched for either more or less stable base-pairing between sister strands of the duplex.

The last source of bias in the distribution of mutations can arise due to founder effects on pooling transformant colonies prior to preparing a heterogeneous population of sequences.

These different sources of bias result in differing frequencies of mutations at different sites. For example, *Figure 1* shows the patterns of mutations obtained on treating a 105 base sequence with hydrazine.

4.3 Considerations in the choice of target fragment(s)

A convenient size of fragment to mutagenize is of the order of 100–200 bp in length, although the techniques described here can be readily adapted for both larger and smaller DNA molecules. For mutagenesis of a fragment more than a few hundred bp in length, it is usually better to create subclones of smaller fragments, preferably fully overlapping. The optimal size of the fragment(s) is determined by several factors that should be considered carefully at an early stage.

(a) The identification of potential mutants is simplified considerably if the length of the fragment of interest is limited such that its sequence can easily be determined from a single set of lanes on a sequencing gel. A

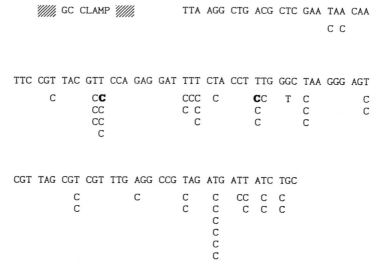

Figure 1. Illustrates the non-random pattern of mutations isolated by the denaturing gel techniques described in the text. The mutated sequences were isolated from gels on the basis of having significantly increased T_m's, following treatment with hydrazine. Only the sequence of the treated strand is shown. The data represent the results of sequencing 60 separate clones, which have yielded 44 mutant sequences, of which one was a double mutation (shown in bold type).

fragment of between one and two hundred base-pairs in length is also ideal for optimum 'single hit' mutagenesis (see Section 4.5).

(b) The smaller the mutagenized fragment the more likely it is to consist of only one melting domain (Section 5). An effective algorithm which predicts the melting behaviour of a molecule of DNA from its primary sequence has been developed by Lerman and Silverstein (9). Application of this algorithm considerably simplifies the decision as to where to subdivide large fragments such that subclones consist of only one melting domain (Section 5).

(c) For a given length of gel and gradient of denaturants, the resolving power of denaturing gels increases with decreasing fragment size. Shorter fragments differing in one base-pair can therefore be more easily separated from each other than can longer fragments with a similar base-pair composition.

(d) The shorter the treated fragment, the greater the degree of mutagenesis which can be applied while still minimizing the probability of multiple 'hits' per DNA molecule.

(e) It is preferable to subject both strands of a sequence to mutagenesis, in order to minimize the effects of the non-random distribution of chemically induced mutations. This is most easily achieved by cloning the target

140

fragment into the mutagenesis vector in both orientations with respect to an M13 origin of replication.

4.4 Vectors for mutagenesis of single-stranded DNA

A pair of suitable vectors, pGC1 and pGC2, developed by Myers *et al.* (8) are schematically illustrated in *Figure 2*. These phagemid vectors contain a suitable 'polylinker' to facilitate cloning, a bacterial selective marker, a bacterial plasmid replication origin, the bacteriophage M13 replication origin, an oligonucleotide primer binding site, and a specialized sequence referred to as a 'GC-clamp' (8). The GC-clamp allows the subsequent separation of mutated sequences from wild-type (Section 5.4).

The use of a pair of vectors, differing only in the orientations of their polylinker sequences, allows a duplex restriction fragment containing the DNA of interest to be cloned in both orientations with respect to the M13 origin. These two vectors will subsequently generate the two complementary strands of the original duplex in single-stranded form, allowing both strands to be exposed to mutagenesis *in vitro*. Single-stranded DNA is prepared as for dideoxy sequencing (*Protocol 1*).

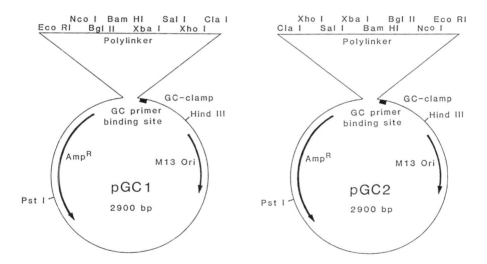

Figure 2. Shows the GC-clamp vectors pGC1 and pGC2 designed by Myers *et al.* (8). The plasmids are derivatives of pBR322 containing the β-lactamase selectable marker and the Co1E1 origin. The plasmids differ only in the orientation of the 65 bp polylinker sequence. This enables sequences to be cloned in either orientation adjacent to the GC-clamp, allowing mutagenesis of both complementary stands of a cloned sequence. Single-stranded DNA can be generated from the bacteriophage M13 origin of replication. Complementary strand synthesis (following mutagenesis or during dideoxy sequencing), can be carried out using an oligonucleotide primer which binds to a site between the polylinker sequence and the GC-clamp. (Adapted from ref. 8.)

4.4.1 Preparation of single-stranded DNA

In order to grow single-stranded, filamentous phage (the source of DNA for mutagenesis) a bacterial strain harbouring an F-factor must be used, the F-pilus being the adsorption target recognized by these phage. The use of an F' factor carrying a selectable marker ensures F' maintenance in the host cell-line. Typically, a F' *lac-proAB* plasmid is maintained in a host bacterial strain deleted for the *lac-pro* region of the *Escherichia coli* chromosome. This can be stored on minimal medium agar without proline, to prevent spontaneous loss of the F'. However, for DNA preparation, the cells should be grown on a rich medium, such as a tryptone-yeast broth. Only rich media allow the proliferative growth of the pili to which the filamentous bacteriophage will subsequently attach. An antibiotic to which the mutagenesis vector confers resistance is also routinely included in this medium. In the case of vectors pGC1 and pGC2 (*Figure 2*) this is ampicillin (typically at a concentration of 100 μg/ml).

Protocol 1. Isolation of single-stranded DNA

1. Grow a 20 ml culture of *E. coli* carrying an F' factor and the recombinant pGC plasmid, to saturation in tryptone-yeast extract broth (TYE, 1) containing 100 μg/ml ampicillin. The formation of bacterial F' pili is inhibited below 34°C and during anaerobic growth thus it is important to ensure that the culture is well-aerated and that the temperature remains between 37° and 39°C, during this period of growth.

2. Add the culture to 800 ml of the same rich medium pre-warmed to 37°C and containing 100 μg/ml ampicillin.

3. Infect the cells with the filamentous helper phage at a m.o.i. of about 50–100 p.f.u. per cell. An ideal phage strain for this purpose is M13rV1 (10) in that it routinely gives high yields of single-stranded DNA.

4. Incubate the culture for 4–5 h at 37°C with vigorous agitation. To ensure adequate aeration use a large volume conical flask on an orbital-type shaker set to rotate to give a rapid swirling action to maximize the surface area of culture in contact with air.

5. Pellet the cells and any cell debris by centrifugation (e.g. Sorvall GSA rotor, 10 000 r.p.m., 10 min).

6. Carefully decant the supernatant to a flask containing 200 ml of a sterile solution of 20% (w/v) polyethylene glycol (mean M_r 6000), 2.5 M NaCl and mix well.

7. Chill the mixture on ice for a minimum of 30 min or store overnight at 4°C.

8. Pellet the phage particles by centrifugation (e.g. Sorvall SS-34, 20 000 r.p.m., 20 min) and gently remove all traces of the supernatant by mild aspiration.

9. Dissolve the pellets in 3–4 ml TE (10 mM Tris–HCl (pH 8.0), 1 mM EDTA).

10. Centrifuge to pellet undissolved material (e.g. Sorvall SS-34 rotor, 10 min) and transfer the supernatant to a fresh tube.

11. Extract the supernatant with an equal volume of 1:1 phenol:chloroform previously equilibrated with the TE. Centrifuge to separate the phases and remove the upper (aqueous) phase, taking care to avoid precipitated material at the phenolic/aqueous interface. Repeat the extraction of the aqueous phase two more times with an equal volume of chloroform to remove traces of dissolved phenol.

12. Add sterile 4 M NaCl to a final concentration of 100 mM and precipitate the DNA by adding 2 vol. of ethanol. Chill at $-70\,^{\circ}$C for 30 min or store at $-20\,^{\circ}$C until required.

13. Centrifuge at $4\,^{\circ}$C (e.g. Sorvall SS-34 rotor, 20 000 r.p.m. for 30 min) to pellet the DNA.

14. Wash the pellet with 70% (v/v) ethanol chilled to $-20\,^{\circ}$C, vortex, and centrifuge for 15 min (e.g. Sorvall SS-34 rotor, 10 000 r.p.m.) at $4\,^{\circ}$C. Repeat using absolute ethanol.

15. Dry the pellet by vacuum desiccation (1 min) and redissolve in 1 ml of TE.

Phenol extractions are conveniently carried out in polythene microcentrifuge tubes. Inclusion of a trace of 8-hydroxyquinoline in the phenol/chloroform mixture not only acts as an anti-oxidant (reducing the formation of oxidation products of phenol which can react destructively with DNA) but also colours the phenolic phase yellow, increasing the contrast between the two phases and considerably aiding the separation of the aqueous phase by clarifying the interface.

Quantitate the DNA by UV absorption spectrometry. For most samples it can be assumed that an absorbance of 1.0 at a wavelength of 260 nm, with a 1-cm light path, corresponds to a concentration of 33 μg/ml of single-stranded DNA, provided no other molecules absorbing in the same region of the spectrum are present. Below an absorbance of one the relationship between concentration and absorbance is close to linear.

The typical yield of single-stranded DNA using this protocol is 2–4 mg which contains plasmid and helper phage DNA in an approximate ratio of 1:1.

4.5 Mutagenic procedures

4.5.1 The statistical limitations on the extent of mutagenic treatment

In general, so that the effects of single alterations in sequence can be observed and studied, one normally wishes to mutagenize only one site within the sequence of interest of each individual DNA molecule. This is directly dependent on the length of fragment being treated. Under the conditions described below in *Protocols 2* and *3* there will be, on average, one 'hit' in every 1500 base-pairs in 10 min. Alternatively, around 10% of fragments of 150 base-pairs in length will be hit in 10 min under these conditions.

However, in order to be able to detect induced mutations efficiently by DNA sequencing, it is desirable that a high proportion of the sequenced molecules contain an altered base-pair. If mutations are distributed at random throughout a molecule, the proportion of DNA molecules changed at a single site, at two sites, at three sites, etc., will follow a Poisson distribution. Consequently, these ideals are mutually conflicting.

In practice, in order to maximize the proportion of molecules mutated at only a single site, the extent of mutagenesis should be adjusted such that between 10 and 20% of molecules experience one or more 'hits'. This, of course, means that 90% of molecules will be unaltered, and 90% of subsequent sequence determinations will yield the wild-type sequence. This problem has been elegantly solved by Fischer, Lerman, Maniatis, and Myers, who have developed a method whereby one- and two-dimensional denaturing gel eletrophoresis techniques can be used to enrich the proportion of mutated molecules in a sample, prior to the determination of their sequences (Section 5).

Protocol 2. The depurination of DNA by mild acid hydrolysis

1. Adjust the concentration of single-stranded DNA to $1 \mu g/\mu l$ with TE (pH 8.0) and place $40 \mu l$ in a microcentrifuge tube.

2. Add $60 \mu l$ of concentrated (18 M) formic acid and mix.

3. Incubate the reaction at room temperature for a time which will result in approximately 10% of the target fragments being hit (see text).

4. Stop the mutagenesis by quenching with $200 \mu l$ 2.5 M sodium acetate (pH 5.5), $100 \mu l$ H_2O, $20 \mu g$ tRNA and 1 ml of cold ethanol. Precipitate the DNA by chilling at $-70°C$ for 30 min.

Protocol 3. Depyrimidination of DNA by hydrazine

1. Adjust the concentration of single-stranded DNA to $1 \mu g/\mu l$ with TE (pH 8.0) and place $40 \mu l$ in a microcentrifuge tube. Do not use a buffer of high ionic strength since the rate of the reaction of hydrazine with thymidine will be reduced.

2. Add $60 \mu l$ of concentrated (12 M) hydrazine and mix.

3. Incubate the reaction at r.t. for the amount of time which will result in approximately 10% of the target fragments receiving a 'hit', assuming the same mutagenesis rate as with formic acid (see *Protocol 2*).

4. Quench the reaction by adding $200\,\mu$l 2.5 M sodium acetate (pH 5.5), $100\,\mu$l H_2O, $20\,\mu$g tRNA and 1 ml of cold ethanol. Precipitate the DNA by incubating at $-70\,^{\circ}$C for 30 min.

4.6 Regeneration of duplex DNA from treated single-stranded DNA

Following mutagenesis of single-stranded DNA, it is necessary to convert this into double-stranded DNA before it can be recloned and transformed into bacterial cells. The majority of DNA polymerases cannot replicate badly damaged DNA *in vitro*. Specifically, most bacterial and viral polymerases terminate chain elongation on encountering a missing base. Fortunately, this is not true of many reverse transcriptases (which lack 'editing' functions). The enzyme usually utilized for the conversion of single-stranded to double-stranded DNA following mutagen treatment is avian myeloblastosis virus DNA polymerase (AMV reverse transcriptase). In common with other DNA polymerases, AMV reverse transcriptase cannot initiate chain synthesis *de novo*. It can only extend the 3'-hydroxy termini of existing chains. A suitable primer (8) is supplied by annealing a synthetic oligonucleotide to a sequence on the treated strand of the vector some distance 3' to the insert. This procedure is identical to that used in dideoxy sequencing, and is described in *Protocol 4*.

Reverse transcriptases in general (including AMV reverse transcriptase) are rather error prone and under *in vitro* conditions misincorporate, on average, one mismatched base-pair per 600 polymerized nucleotides under the conditions described in *Protocol 4* (R. K. Booth, unpublished data). During regeneration of a DNA duplex of 150 base-pairs this might result in one in four DNA molecules carrying reverse transcriptase generated mutations in addition to those induced by prior chemical treatment of the nucleic acid. In fact this situation is both complicated by and simplified by the biology of the bacterial cell. Misincorporation of the wrong base is detected by the cell following transformation, and is subsequently corrected by the bacterial DNA repair systems.

If no further information was available to the bacterium, either strand might be corrected, reducing the observed mutation rate caused by reverse transcriptase to one in eight for a fragment of 150 base-pairs in length. In fact, the cell can determine which strand is the newly synthesized (reverse transcriptase generated) strand and correct this preferentially. DNA methylation of adenosine and cytidine bases during the original synthesis *in vivo* of single-stranded DNA labels this strand as that **not** to be corrected. Consequently, while errors are introduced during the *in vitro* polymerization reaction with reverse transcriptase, the majority of these mutations are subsequently removed *in vivo* following transformation. Of course, this does not apply to mutations induced by hydrazine or mild acid mutagenesis, where the newly incorporated base is not paired with a corresponding base in the DNA duplex.

Protocol 4. Second-strand synthesis using AMV reverse transcriptase

1. Remove traces of the mutagen from the precipitate of chemically treated DNA from *Protocols 2* and *3* by redissolving in 10 mM Tris–HCl (pH 7.5), 100 mM NaCl, 1 mM EDTA, containing 20 µg/ml DNase free nucleic acid as carrier. (Convenient commercial sources of carrier nucleic acid are the high quality tRNA preparations available from a number of manufacturers.) Reprecipitate by adding 2 vol. of ethanol and chill at −70°C for 30 min.

2. Repeat step 1 at least once more to ensure the thorough removal of formic acid or hydrazine.

3. Wash the pellet with 1 ml 70% (v/v) ethanol and allow the pellet to dry. Dissolve in 80 µl TE (pH 7.5). Add 10 µl of 70 mM Tris–HCl (pH 7.5), 70 mM $MgCl_2$, 0.5 M NaCl, 20 mM dithiothreitol (or dithioerythritol), and 40 pmol of oligonucleotide primer complementary to the cloning/sequencing vector being used (1).

4. Heat the sample to 75°C for 5 min. Transfer to 40°C and incubate for 15 min to anneal the primer to its complementary sequence on the treated DNA.

5. Add 8 µl of dNTP solution containing 2 mM of each deoxynucleotide triphosphate to give a final reaction volume of 100 µl.

6. Add 30–40 units of AMV reverse transcriptase and incubate at 40°C for 1 h. Terminate the reaction by placing in wet ice.

7. Extract with an equal volume of 1:1 phenol:chloroform, equilibrated with TE (pH 8.0), and containing approximately 0.1% (w/v) 8-hydroxyquinoline. Remove the aqueous phase and re-extract with an equal volume of chloroform. Precipitate the DNA with two volumes of ethanol at −70°C for 30 min.

4.7 Amplification and enrichment of treated sequences

Before it is possible to enrich for mutated sequences by gradient denaturing gel electrophoresis, it is necessary to amplify the mutated sequences in order to generate enough material to visualize on a gel. Following regeneration of double-stranded DNA (see *Protocol 4*), the sequence of interest is excised, separated from the rest of the vector by electrophoresis, cloned into a suitable sequencing vector, and transformed into competent bacterial cells. These are plated out at densities of several hundred colonies per plate. After incubation, the colonies are pooled and grown together to give a culture of mixed cells, from which plasmid DNA is prepared. This DNA is a heterogeneous preparation, isolated from a polymorphic population of cells containing clones of mutated and unmutated sequences. These procedures are described in detail in *Protocol 5*.

The fragment of interest is then excised and the mixed population of DNA molecules enriched for altered sequences by one-dimensional denaturing gradient gel electrophoresis (Section 5.2.3), prior to recloning into a suitable sequencing vector and determination of the sequences of as many clones as required.

Protocol 5. Preparation of mutant-containing plasmid DNA

1. Digest the DNA from *Protocol 4* to excise the fragment of interest from the now damaged vector by using appropriate restriction enzymes in a total reaction volume of $200\,\mu l$.

2. Add $5\,\mu l$ of 4 M NaCl and precipitate the digested DNA with two volumes of ethanol. Resuspend the pellet in a small volume of non-denaturing gel loading buffer (10 mM Tris–borate, 2.5 mM EDTA (pH 8.0; TBE) containing 10% (v/v) glycerol, 0.05% (w/v) xylene cyanol and 0.05% (w/v) Bromophenol Blue) and to this add approximately $5\,\mu g$ of RNase A (pretreated at 100°C to inactivate contaminating DNase) to remove the tRNA carrier.

3. Separate the mutagenized fragments from the vector DNA by electrophoresis in a polyacrylamide gel (typically 8% (w/v)) made up in TBE. Soak the gel in a solution of ethidium bromide ($1\,\mu g/ml$) and visualize the DNA under long-wavelength UV light (~310 nm). A bright band of ethidium staining material corresponding to the mutagenized double-stranded fragment of interest will usually stand out against a background smear. However, it is advisable to run purified DNA of an untreated sample of the same DNA fragment in an adjacent lane as a marker for the position of the target DNA.

4. Excise the section of the gel containing the DNA of interest and elute the DNA by the 'crush-soak' method (11). Crush the gel fragment into 0.5 ml of elution buffer (0.5 M ammonium acetate, 1 mM EDTA, 0.1% (w/v) SDS and $20\,\mu g$ tRNA carrier). Incubate overnight at 37°C with shaking.

5. Centrifuge for 1–2 min at $10000\,g$. Remove the supernatant and filter through siliconized glass fibre to remove the smaller pieces of polyacrylamide. Wash the pellet of polyacrylamide with $200\,\mu l$ of the same buffer, filter, and add to the supernatant.

6. Precipitate the DNA by adding two volumes of ethanol and chill for 30 min at −70°C. Wash with a 70% (v/v) solution of ethanol to remove traces of SDS. Dissolve the pellet in $20\,\mu l$ of TE (pH 8.0).

Protocol 5. *continued*

7. Remove a portion of this mutagen-treated double-stranded DNA and ligate with untreated plasmid vector, cut with appropriate restriction en-donucleases. Use this to transform competent bacterial cells (1).

8. Pool the transformant colonies by pouring 3 ml of TYE medium on to each plate and suspending the cells on the surface of the plate using a sterile glass spreader. Use this suspension[a] to inoculate 500 ml of fresh bacterial growth medium and grow at 37°C for 2–4 h.

9. Prepare purified plasmid DNA from the pooled colonies by a suitable method (1).

___

[a] It is surprisingly easy to suspend too many cells at this point, such that little growth occurs in the rich medium due to the high starting cell density. The cells should be allowed to grow through at least four doublings between inoculation and harvesting.

___

It is advisable to obtain as large a number of transformed colonies as possible. The number of base-pair substitutions subsequently detected will be proportional to the number of transformed colonies obtained at this stage. Consequently, the number of unique mutations detected will increase with the number of transformed colonies, asymptotically approaching the theoretical maximum of all possible changes at all possible sites.

The choice of plasmid purification technique employed in *Protocol 5* is a matter of individual preference. We prepare our DNA from caesium chloride gradients as this technique is efficient and capable of separating large amounts of plasmid DNA in a single step.

It is important to note that DNA prepared from cultures inoculated from different plates of transformant colonies will give rise to different sets of mutations. Even from a single plate of initial transformant colonies, different inocula from the same 'washing' often give rise to a different distribution of mutations in the subsequent plasmid preparation due to the 'founder effect' (Section 4.2).

4.8 Nature of the mutations generated

The depurination and depyrimidination methods described in Section 4.5, will, in a large enough population of treated DNA molecules, generate all possible single base-pair changes (although, due to differences in reaction rates between different bases, not at the same frequencies). However, it should be noted that if only one 'hit' per molecule is obtained (usually the ideal mutagenesis dosage, see Section 4.5), this is not the same as generating all possible amino acid changes in a coding sequence. First, as any particular base can be changed only into one of the other three, only nine other codons are accessible as single base changes to any particular triplet. When one takes into account the three-chain termination codons, this means that 34 of the 61 remaining codons can be changed into triplets encoding nine other amino acids, 25 into eight others, and two (UUA and UCA) into only seven other amino acid coding triplets.

Second, most amino acids are encoded by more than one codon. In fact only two amino acids are encoded by a single codon, whilst three amino acids are encoded by six codons and a further six by four. The pattern of this degeneracy is not such that the codons are assigned to amino acids at random. Rather, related codons encode chemically similar amino acids, and where more than one codon is used for an amino acid, those codons differ from the other codons encoding that same amino acid only in their third base. The overall consequence is that many changes in nucleotide sequence do not result in changes in the amino acid sequence.

This limitation of the spectrum of possible amino acid substitutions by a single base change is not a product of the chemistry of these procedures, but is a natural effect arising from the pattern of degeneracy in the genetic code. Examination of the code reveals that if all the codons for an amino acid are used with equal frequency, 24% of single base-pair changes in an amino acid coding sequence will not affect the sequence of the protein encoded, whilst a further 14% will result in the substitution of an amino acid by one chemically rather similar (alanine by valine, or serine by cysteine, for example). In fact, the observed situation is more complex, partly because where several codons encode an amino acid, some are used more frequently than others, and also because the amino acids found most commonly in proteins are generally those which are encoded by the most codons. This results in random mutagenesis *in vitro* generating an even higher proportion of 'translationally silent' base-pair changes than would be expected from consideration of the code alone.

5. Electrophoretic techniques allowing the enrichment of mutants

In most cases it is now possible to separate, or partly separate, duplex molecules of DNA of a few hundred base-pairs (or less) in length, from sister molecules differing in sequence by as little as a single base-pair. Techniques which successfully exploit a combination of the denaturation of duplex DNA by heat and by chemical interaction, together with the resolving power of electrophoretic separation, have been developed and refined over the past decade (14, 15, 16). Myers has pioneered the application of these separation techniques to mutagenesis *in vitro*. The result of this work is a method whereby DNA subjected to mutagenesis can be enriched for altered sequences. This combination of techniques, which has become known as denaturing gradient gel electrophoresis (DGGE), exploits the slight differences in the 'melting' properties of altered sequences, to separate mutant molecules from the non-mutant (wild-type) sequence. Its application has opened the way for the practical screening for mutants by sequence determination alone whilst still allowing the extent of mutagenesis to be limited to one 'hit' per DNA molecule (see Section 4.5).

5.1 Theory of denaturing gradient gel electrophoresis (DGGE)

If the wild-type fragment is electrophoresed through a polyacrylamide gel containing a gradient of increasing concentration(s) of denaturant(s), a point is reached at which the duplex DNA will dissociate into single strands. This can be thought of as equivalent to the melting temperature of DNA (T_m) denatured by heat alone. The effect of chemical denaturants is to lower the actual denaturation temperature by changing the nature of the solvent such that single-stranded DNA is thermodynamically more favoured than in more physiologically realistic buffers.

Duplex DNA, constrained in the conformation it adopts by two antagonistic covalent 'backbone' bonds at any particular point, behaves much as a rigid cylinder during electrophoresis. Alignment of molecules within an electrophoretic gel allows DNA to migrate rapidly through the gel, relative to other molecules of comparable molecular weight. In contrast, the increased flexibility of single-stranded DNA (which behaves largely as a 'random coil') results in the molecules becoming entangled within the structure of the gel itself, dramatically reducing their mobility. This change in mobility on dissociation into single-strands is so dramatic that if a molecule of duplex DNA can be persuaded to dissociate into single-strands during its migration through a gel, it will be perceived as effectively having stopped migrating through the gel at that point.

5.1.1 DNA melting domains

Duplex DNA does not melt base by base, but in domains of dozens of bases at a time. The T_m of such a melting domain is determined by the energy of the hydrogen-bonding and the 'stacking' energies which hold the two strands together; that is, by its sequence. Sequences differing by as little as one base-pair can show altered T_m's sufficient to allow separation of such sequences on a polyacrylamide gel. The technique is sensitive enough to distinguish not only between transition type mutations (which have an altered number of hydrogen bonds between the two strands of the molecule) but also between C-G to G-C and A-T to T-A transversion mutations, exploiting the differences in the 'hydrophobic' stacking energies between adjacent base-pairs.

In practice, separation is carried out by electrophoresing the mutated duplex DNA into a continuously increasing concentration of chemical denaturant. In such gels the wild-type DNA migrates until it reaches the concentration at which it dissociates into single-strands. On visualization of the

DNA, the wild-type appears as a single band with mutant DNA molecules visible as a near continuum of closely spaced bands above and below the wild-type band.

Fragments of duplex DNA larger than a few hundred base-pairs in length, and usually much less, rarely melt simply as a single domain. As each domain dissociates, a reduction in mobility occurs. With such molecules, mutations in the domain with the lowest T_m are detected with the greatest resolution, as the melting point of subsequent domains has a reduced effect on the mobility of the already retarded molecule. The proportion of mutants resolvable in such cases can be maximized at the initial cloning stage by dividing larger DNA fragments into sizes which correspond to one melting domain.

5.1.2 G–C clamps

The drawback of using denaturing gels to separate DNA molecules which are subsequently to be cloned is that one might usually expect only to recover isolated single-stranded molecules from the gel. If the DNA is to be subsequently recloned the regeneration of duplex molecules suitable for ligation into a plasmid vector presents a problem. This has been overcome by Myers *et al.*, by the use of what is termed a 'GC-clamp' (12, 13). This is a region of DNA rich in G-C base-pairs, incorporated into the original single-strand vector between the restriction sites used to excise the mutagenized fragment (see *Figure 2*).

Since guanosine and cytidine are bonded by three hydrogen bonds, G–C-rich regions dissociate into single-strands at higher temperatures or higher denaturant concentrations than sequences with a more typical composition. In practice, attaching the mutagenized fragment to a G–C-rich 'clamp' results in the melting of the region of interest, with a concomitant retardation of the nucleic acid's mobility within a gel, while the GC-clamp remains base-paired, holding the two otherwise dissociated strands of mutagenized sequence together. These reanneal on excision from the denaturing gradient gel, thus allowing recloning into a suitable vector.

The analysis of a large number of different mutations in the β-globin promoter, together with theoretical calculations, indicate that approximately 95% of all possible single-base substitutions should be separable when attached to a GC-clamp (13). A pair of suitable single-strand generating cloning vectors, each allowing the excision of a DNA fragment attached to a GC-clamp in one of the two possible orientations, are pGC1 and pGC2 (see *Figure 2*).

5.2 Apparatus

Several options are available for DGGE systems but two considerations must be borne in mind when deciding upon the construction, adaptation or purchase of suitable apparatus for running denaturing gradient gels.

5.2.1 Running buffer

The gel is run (see *Protocol 7*) at a temperature of 60 °C which is close to the melting temperature of most naturally occurring DNA. This is most easily achieved by running the gels submerged in a bath of heated electrode buffer. The equipment consists of a tank (we use a modified aquarium, purchased from a domestic pet shop) full of electrophoresis buffer in which the gel apparatus is almost completely submerged. The walls of the upper reservoir of the gel apparatus should be above the level of the buffer in the tank, thereby separating the upper electrode reservoir from the lower; in this case the rest of the tank. The buffer in the tank is circulated within the tank to maintain an even temperature, which is regulated with a thermostatically controlled heater. In addition, buffer is circulated from the tank to the upper reservoir to maintain the pH and ionic strength in the upper electrode reservoir during the sometimes long periods of electrophoresis required.

For reasons of safety, as well as reproducible behaviour of the apparatus, it is important that the path of buffer through the pump does not form a path for electric current. This can easily be achieved by allowing the buffer to drip from the pump outlet into the upper (cathode) chamber of the apparatus. The anodic chamber of the apparatus can be grounded to bring the buffer tank and pump down to zero voltage relative to the operator. A small hole can be placed in the wall of the upper reservoir to allow the overflow of the recirculating buffer to drip into the tank. Graphite or platinum electrodes can either be built-in to the equipment or kept separate and be appropriately positioned when running the gels. Gels are typically run at 10 V/cm.

5.2.2 Gel casting

If perpendicular gels are to be run (Section 5.3.1; *Protocol 6*) it must be possible to cast the gels rotated through 90 degrees relative to the direction in which they are to be run. This allows a gradient gel to be cast on its side so that the variation in the concentration of denaturant(s) changes **across** a gel when it is subsequently electrophoresed vertically. This is most easily achieved by casting the gel through a small hole positioned on one side of the frame about 5 cm up from the base. It is also necessary to be able to temporarily seal what will be the open top of the gel. The seal is removed once the acrylamide has polymerized. Details of the construction of this equipment, together with the addresses of commercial suppliers, are given by the original developers of this gel system (14, 15, 16). A photograph of apparatus suitable for pouring and running gels is shown in *Figure 3*.

A recent development, not dealt with here, allows denaturing gels to be run with a temperature gradient increasing down the gel so dispensing with the need for chemical denaturants (17, 18). In theory, the resolution of the system can be increased still further by a combination of chemical and thermal denaturation, introducing a second control parameter into the separation.

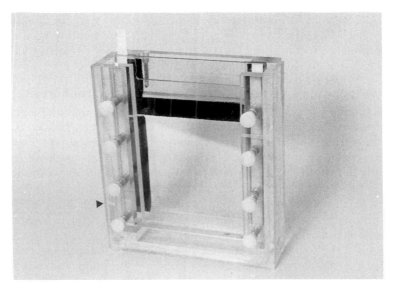

Figure 3. The acrylic frame is an example of a convenient gel support which can be used for casting denaturing gradient gels. The hole at the lower left side of the gel apparatus is used for pouring gels with a horizontal gradient of denaturants with the apparatus rotated through 90 degrees and the hole uppermost (Section 5.3.1). For electrophoresis the acrylic frame containing the gel is placed into a tank of heated electrophoresis buffer and the cathode positioned in the upper compartment of the gel apparatus. The anode is placed in the electrophoresis tank.

The potential of this approach has yet to be fully explored. The temperature gradient gel equipment can be constructed (17, 18) but is also now available commercially from Diagen GmbH.

5.3 Determination of the optimum conditions for DGGE

Before DGGE can be used to separate mutant fragments from wild-type DNA (*Protocol 9*), it is necessary to know the range of denaturant concentrations and time of electrophoresis which will result in all the molecules of interest reaching their denaturation points. Consequently, the melting behaviour of the wild-type fragment linked to the GC-clamp must first be determined. There are two ways of doing this.

(a) An algorithm has been developed by Lerman and Silverstein (9), to analyse the DNA sequence of the fragment to be mutated if this is known. This algorithm, available in the form of a computer program written for an IBM PC or compatible computer, enables the rapid theoretical prediction of the positions and T_m's of melting domains along the fragment. This allows the optimal slope of the denaturant gradient concentrations and duration of electrophoresis to be calculated.

(b) Alternatively the melting behaviour and mobility of the fragment can be determined empirically using perpendicular denaturing gradient gel electrophoresis (Section 5.3.1) and a 'time schedule' gel respectively (Section 5.3.2).

5.3.1 Perpendicular gradient denaturing gel electrophoresis

This technique is used to determine how many melting domains a particular DNA fragment contains, and to measure their denaturant induced T_m's at a given temperature. The DNA fragment is electrophoresed (see *Protocol 7*) vertically through a gel with a horizontal denaturant gradient prepared according to *Protocol 6*. At any point across the gel, the mobility of the DNA, and thus the distance travelled vertically during electrophoresis, is dependent on the concentration of denaturant. In particular, the mobility of the fragment within the gel is drastically altered by whether the denaturant concentration at a particular point across the gel is above or below that required to dissociate the two strands of the DNA sequence under investigation.

Following staining of the gel with ethidium bromide and visualization with long-wavelength UV light, the DNA on the gel traces out a steep, sigmoidal 'melting curve', which can be calibrated from the known denaturant gradient across the gel. *Figure 4* shows a typical melting-point curve obtained from a perpendicular gel where the fragment of DNA attached to the GC-clamp consists of a single melting domain. In the case of DNA molecules with more than one melting domain, several transitions are seen in the DNA curve.

Protocol 6. Preparation of perpendicular denaturing gels

1. Place the gel plates in the apparatus (see *Figure 3*) and position 0.6-mm spacers at both sides of the plates and a wide spacer at the top protruding above the glass plates. The left spacer should be raised 3–4 cm so that there is a gap at the bottom of the plates in alignment with the hole in the side of the frame. Seal the junction between the right and top spacer with an inert grease such as silicone vacuum, and seal the outside of the right spacer with a molten solution of 1.5% agarose in electrophoresis buffer (40 mM Tris–HCl, 20 mM sodium acetate and 1 mM EDTA, adjusted to pH 7.4 with acetic acid). The lower end of the gel plates can be sealed by pouring 1.5% agarose into the bottom of the frame and allowing it to set.

2. Rotate the gel plates through 90 degrees with the hole uppermost.

3. Prepare acrylamide stock solutions of **'0% denaturants'** (6.5% (w/v) acrylamide monomer solution (37.5:1 acrylamide:bisacrylamide) in electrophoresis buffer) and **'80% denaturants'** (6.5% (w/v) acrylamide (37.5:1 acrylamide: bisacrylamide), 5.6 M ultrapure urea, 32% (w/v) deionized formamide in electrophoresis buffer).

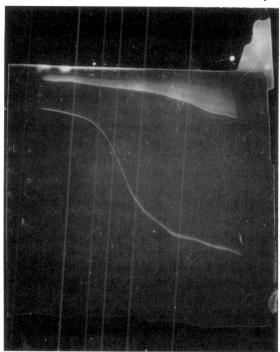

Figure 4. A typical melting-point curve obtained by perpendicular gradient gel electrophoresis. On this particular gel, the gradient decreases from 80% to 10% of denaturants from left to right, giving a T_m of approximately 47% denaturants at the 60°C temperature at which this gel was run (see Section 5.3.1). The brightly fluorescing material with slow mobility, seen close to the top of this gel, is the vector from which the fragment of interest has been excised.

4. Place acrylamide containing 80% denaturants in the outlet chamber of a gradient maker and an equal volume of acrylamide containing 10% denaturants in the other chamber. Take care to ensure that no air bubbles are trapped in the channel between the two chambers of the gradient maker. A partial blockage can easily go unnoticed and can lead to a non-linear gradient being created, and consequently an incorrect estimate of the T_m of the fragment.

5. Keep the solution in the outlet chamber of the gradient maker stirring well (on a magnetic stirrer) during the pouring of the gel. Place tubing from the exit of the gradient maker *via* a peristaltic pump to a hypodermic needle positioned between the glass plates. Adjust the speed of the pump such that the acrylamide solution will run into the apparatus as a constant steady stream, thereby forming an even gradient.

6. Immediately prior to casting the gel, add 1/200th of the gel volume of 20%

Protocol 6. *continued*

(w/v) ammonium persulphate and 1/1000th of the gel volume of N,N,N',N'-tetramethylethylenediamine (TEMED) to each chamber.

7. Switch on the pump and stirrer, and open the stopcock between the chambers of the gradient maker. Check that the solutions are being mixed, which can be seen as rapidly dispersing Schlieren fringes within the outlet chamber, and that there are no blockages, such as air bubbles, in the channel connecting the chambers. The gel should be poured smoothly with the minimum of turbulence in about 3–5 min.

8. Allow the gel to polymerize fully for at least 30 min after it solidifies. Rotate the gel so that it stands upright and remove the spacer used to seal the top of the apparatus during the pouring of the gel. This should create a single well across the entire width of the gel. Sometimes a channel forms along the left side of the gel due to leakage or shrinkage of the gel. If so, this can be sealed with molten 1.5% (w/v) agarose in electrophoresis buffer.

Occasionally problems can arise during the pouring of perpendicular gradient gels due to leakage of the gel solution between the well former and the body of the apparatus. This usually happens when the gel is almost poured, and leakage usually occurs low down in the apparatus where the hydrostatic pressure is highest. One solution to this is to increase the concentration of TEMED in the solution so that polymerization is rapid. To avoid polymerization in the gradient maker and tubing while the gel is being poured, it is necessary to use a gradient of TEMED with the higher concentration in the solution containing the higher concentrations of denaturants. We suggest a three to one concentration gradient of TEMED. The persulphate should be made up fresh every two to three days and stored at 4°C or −20°C. Even dry persulphate is somewhat unstable and decayed persulphate is the most common source of a failure of acrylamide gels to polymerize.

If problems are experienced due to polymerization in the gradient maker before completion of gel casting, the rate of polymerization can be slowed down by pre-chilling both the solutions of acrylamide and the gradient maker itself to 0°C. This slows the rate of polymerization in the gradient maker, while still allowing the gel to polymerize in a convenient time. The solutions warm rapidly as the gel is cast due to the relatively high heat capacity of the glass plates of the apparatus.

Although it is assumed here that a gradient maker will be used to form the gradient, we often find that we obtain better reproducibility between gels by forming the gradient using differential pumping. The volume of each of the acrylamide solutions used should be exactly equal to half the final volume of the gel.

Protocol 7. Electrophoresis

1. Submerge the gel apparatus in the tank containing preheated electrophoresis buffer (see *Protocol 6*, step 1) and connect the pump to recirculate the buffer from the lower to the upper reservoir (see Section 5.2.1). Fill the upper reservoir with electrophoresis buffer.
2. Load the DNA sample evenly across the full width of the gel in $200\,\mu l$ of loading buffer (electrophoresis buffer containing 10% (v/v) glycerol, 0.05% (w/v) Bromophenol Blue and 0.05% (w/v) xylene cyanol). The DNA sample is usually a restriction digestion in which the fragment attached to the GC-clamp has been excised from about $10-20\,\mu g$ of the plasmid containing **wild-type** fragment.
3. Electrophorese the sample at 10 V/cm for sufficient time to allow unmelted DNA to reach the bottom of the gel. For a fragment of 150 basepairs and a gel of 15 cm in length this is around 8 h.
4. Stain the gel by soaking in a $1\,\mu g/ml$ solution of ethidium bromide in electrophoresis buffer and destain by washing in TE for 10–15 min.
5. Visualize the DNA under long-wavelength UV illumination. Position a graduated ruler along the bottom of the gel and photograph the gel and ruler. This allows subsequent calibration of the denaturant concentrations across the gel. Consequently, the midpoint of each melting transition can be measured from the photograph.

The optimum gradient of denaturants for use in preparative gels can be calculated as the concentration range plus and minus 15% centred on the measured T_m of the fragment, estimated from the perpendicular gel photograph, or from the computer algorithm. Mutants should be separated in the orientation, with respect to the GC-clamp, that shows the simplest melting behaviour. Not infrequently, a given DNA sequence will melt as a single domain when attached to a GC-clamp in one orientation, but as two domains when attached in the opposite orientation.

If the fragment melts as more than one domain when attached to the GC-clamp in both orientations, it is possible to detect mutants in each of the melting domains on a single preparative gel. In this case the range of concentrations chosen should be from 15% less than the T_m of the least stable domain, to 15% more than that of the most stable domain. In such a case, it will not be possible to resolve the domains with the higher T_m's to as great a resolution as those in the domain(s) with the lower $T_m(s)$. Maximum resolution can be obtained by using as many different gels as there are melting domains in the fragment, each with a range of denaturant concentrations centred around each T_m. However, it is probably counter-productive to attempt to enrich for mutations in fragments which melt as more than two domains.

5.3.2 'Time-schedule' gels

These gels contain a gradient of denaturant(s) spanning a range a little wider than that needed to dissociate the two strands of the fragment. They are used to determine the time needed for separation or enrichment of mutated molecules from the wild-type on a preparative gel. The fragment of interest is electrophoresed in parallel lanes, as in conventional polyacrylamide gel electrophoresis, through a gel containing a gradient of denaturants. However, unlike the perpendicular gradient gels described above, the gradient of denaturant(s) now runs vertically, increasing from top to bottom.

Time-schedule gels differ from conventional polyacrylamide gradient gels in that samples in different lanes are electrophoresed for different times as outlined in *Protocol 8*. Subsequent comparison of the mobility of fragments in different lanes allows the accurate determination of when a particular fragment melts. On melting, the DNA in a particular lane will be perceived to have almost stopped migrating, when compared to unmelted DNA in adjacent lanes. This differential electrophoresis time is achieved by starting the electrophoresis with only a single lane loaded. At intervals following the start of the electrophoretic run, the apparatus is switched off and another lane loaded with the sample. At the end of the run samples that have dissociated will have almost stopped migrating through the gel at a point in the gel corresponding to their dissociation concentration of denaturant(s).

Protocol 8. Analysis of time-schedule gels

1. Digest the wild-type fragment attached to the GC-clamp from a suitable plasmid using the appropriate restriction enzymes.

2. Load a DNA sample into the first well of a denaturing gel cast with the optimum denaturant range plus and minus 15%. Load subsequent samples in adjacent lanes at various intervals following the commencement of electrophoresis. Up to 10 h electrophoresis is usually sufficient but depends on both the length and sequence of the fragment under investigation and as long as 24 h may be needed.

3. Stain the gel with ethidium bromide (*Protocol 7*, step 4) and examine under UV illumination.

The time required to achieve separation of mutants from the wild-type is calculated as the time which elapsed from loading the last sample to have dissociated (and therefore stopped migration through the gel) plus two extra hours to allow for the complete separation of mutants with a slightly higher T_m.

5.4 Preparative denaturing gels

These gels are used to prepare DNA enriched for mutations. This DNA is then excised from the gel and recloned into a suitable sequencing vector. Altered DNA sequences are then screened for by DNA sequencing.

Protocol 9. Preparative denaturing gel electrophoresis

1. Digest, with appropriate restriction endonucleases, approximately $10\,\mu g$ of the population of plasmids containing the chemically induced mutations (*Protocol 5* step 9) to release the fragment attached to the GC-clamp.

2. Precipitate the DNA with ethanol and redissolve in loading buffer (*Protocol 7*, step 2).

3. Using a suitable 'well comb' (5- to 8-mm wide wells are suitable), cast a denaturing gradient gel spanning the optimal concentration range (see Section 5.3) and such that the concentration of denaturants increases vertically down the gel (i.e. in the direction of electrophoresis).

4. Allow the gel to polymerize fully and immerse the apparatus in the tank of buffer, pre-heated to 60 °C. Load $4-6\,\mu g$ of the DNA sample into a well of the gel and electrophorese for the time determined from the time-schedule gel[a] (Section 5.3.2).

5. Stain the gel with ethidium bromide (*Protocol 5*, step 3), destain briefly and place on a transilluminator. The wild-type DNA will have electrophoresed as a single band and mutant DNA molecules should be visible as a near continuum of closely spaced bands both above and below the wild-type band. This mutant DNA may be apparent just as a faint smear above and below the wild-type band.

6. Excise the portions of the gel containing the mutant DNA, taking care to avoid the wild-type band. Elute the DNA from the polyacrylamide (*Protocol 5*, step 4).

7. Ligate the eluted fragment into a sequencing vector (for example, pGC1/pGC2). Successful ligation depends upon the removal of all traces of formamide and urea by repeatedly dissolving in TE and reprecipitating with 70% (v/v) ethanol at -20 °C in the presence of $100\,\text{mM}$ NaCl.

---

[a] On denaturation of the DNA duplex in the gel the decrease in mobility is so dramatic that it is most unusual for DNA to electrophorese out of the bottom of the gel. Consequently, it is better to overestimate rather than underestimate the electrophoresis time required for a sample to reach its denaturation point.

The preparative gels which are most successful in enriching for mutant sequences are those loaded with just sufficient nucleic acid to visualize and elute. Excessive loading leads to contamination of the mutant regions of the gel above the wild-type band with unmutated DNA. Obviously, mutant DNA can be isolated from below the wild-type band no matter how overloaded the lane. It may prove necessary to try different loadings in order to successfully elute a population of molecules enriched for mutants from above the wild-type band.

5.5 Analytical denaturing gradient gel electrophoresis

If the proportion of mutant molecules eluted from the preparative gel is reasonably high, they can be screened for mutations by sequencing (1, 19). At least two-thirds of the sequences determined should be mutant sequences. The proportion of molecules which are mutant can be estimated using analytical denaturing gradient gel electrophoresis described in *Protocol 10*. A decision as to whether to proceed with sequence determinations can then be made.

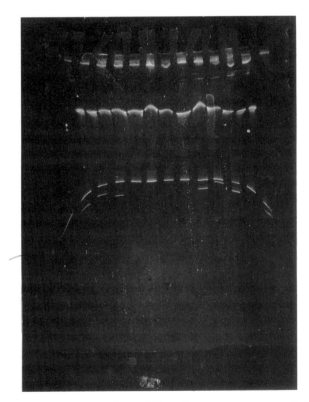

Figure 5. Shows the differences in mobility of mutants on an analytical denaturing gradient gel (Section 5.3.4). The samples are GC-clamped fragments of putative mutant clones electrophoresed together with the corresponding wild-type fragment through a gel of increasing denaturant concentration. In lanes containing mutant sequences, two bands can be seen corresponding to the wild-type fragment and the mutated fragment. In this example, all mutations create sequences with a higher T_m than the wild-type, and therefore greater mobility within a denaturing gradient gel.

Protocol 10. Analytical screening of clones

1. Prepare the plasmid DNA containing the recloned mutant sequence using a rapid 'mini' preparative technique (for example see ref. 1).

2. Release the target fragment attached to the GC-clamp using the appropriate restriction enzymes.

3. Mix a sample of the digested potentially mutant plasmid DNA with a digest containing the corresponding wild-type fragment and electrophorese the combined sample on a one dimensional gel as for preparative electrophoresis (Section 5.3.3).

Mutant sequences can be distinguished by the presence of two bands of differing mobility in one lane of the gel as shown in *Figure 5*. The mutants obtained can be characterized by sequencing.

References

1. Perbal, B. (1988). *A Practical Guide to Molecular Cloning*. John Wiley, New York.
2. Auerbach, C. (1977). *Mutation Research: Problems, Results and Perspectives*. Chapman & Hall, London.
3. Brown, D. M., MacNaught, A. D., and Schell, P. D. (1966). *Biochemical Biophysics Research Communications*, **24**, 967.
4. Shapiro, H. S. and Chargaff, E. (1957). *Biochimica et Biophysica Acta*, **26**, 608.
5. Schuster, H. (1960). *Biochemical Biophysical Research Communications*, **2**, 320.
6. Lawley, P. D. (1967). *Journal of Molecular Biology*, **24**, 75.
7. Shapiro, R., Braverman, B., Louis, J. B., and Servis, R. E. (1973). *Journal of Biological Chemistry*, **248**, 4060.
8. Myers, R. M., Lerman, L. S., and Maniatis, T. (1985). *Science*, **229**, 242.
9. Lerman, L. S. and Silverstein, K. (1987). In *Methods in Enzymology* (ed. R. Wu), Vol. 155, pp. 482–501. Academic Press, London, and New York.
10. Levinson, A., Silver, D., and Seed, B. (1984). *Journal of Molecular and Applied Genetics*, **2**, 507.
11. Maxam, A. and Gilbert, W. (1980). In *Methods in Enzymology* (ed. L. Grossman and K. Moldave), Vol. 65, pp. 499–560. Academic Press, London.
12. Myers, R. M., Fischer, S. G., Maniatis, T., and Lerman, L. S. (1985). *Nucleic Acids Research*, **13**, 3111.
13. Myers, R. M., Fischer, S. G., Maniatis, T., and Lerman, L. S. (1985) *Nucleic Acids Research*, **13**, 3131.
14. Myers, R. M., Maniatis, T., and Lerman, L. S. (1987). In *Methods in Enzymology* (ed. R. Wu), Vol. 155, pp. 501–27. Academic Press, London.
15. Fischer, S. G. and Lerman, L. S. (1979). In *Methods in Enzymology* (ed. R. Wu), Vol. 68, pp. 183–91. Academic Press, London.
16. Myers, R. H., Sheffield, V. C., and Cox, D. R. (1988). In *Genome Analysis—A Practical Approach* (ed. K. E. Davies), pp. 95–139. IRL Press, Oxford.

17. Rosenbaum, V. and Riesner, D. (1987). *Biophysical Chemistry*, **26,** 235.
18. Riesner, D., Steger, G., Zimmat, R., Owens, R. A., Wagenhöfer, M., Hillen, W., Vollbach, S., and Henco, K. (1989). *Electrophoresis*, **10,** 377.
19. Sanger, F., Nicklen, S., and Coulson, A. (1977). *Proceedings of the National Academy of Sciences of the USA*, **74,** 5463.

8

An enzymatic method for the complete mutagenesis of genes

JONATHAN KNOWLES and PÄIVI LEHTOVAARA

1. Introduction

Site-directed mutagenesis has contributed considerably to our understanding of the relationship between protein structure and function. However, since site-directed mutagenesis is laborious, at least with regard to all the possible mutants that should be examined for a full understanding of any particular aspect, it is important that other more efficient methods for the investigation of protein structure by mutagenesis should be available. More importantly, site directed mutagenesis experiments are limited by the imagination of the investigator which has proved in recent years to be often too limited to understand the complexities of proteins.

A method that is becoming increasingly popular is the selection of specific mutants from a complete population of mutant proteins that contains all of the possible single amino acid changes. In this type of experiment the mutants obtained have been selected for function or lack of it, without regard to position, and therefore it is possible to obtain much information that cannot be obtained by any amount of structure gazing. There are at least two problems associated with this type of approach. The first is the development of a specific selection technique that reveals the mutants of interest. The second is the efficient generation of the library of single amino acid mutants.

It is important that the complete library is as small as possible so that even elaborate screening procedures can be used. An ideally efficient random mutagenesis method therefore should only mutagenize within the region of interest. Each nucleotide position should ideally be mutagenized with the same probability, and each base changed to the three other alternatives in equal ratio. The wild-type background should be non-existent.

1.1 Other random mutagenesis methods

The previously described random *in vitro* mutagenesis methods include use of various mutagenic chemicals (1, 2, 3; see Chapter 7). These methods are, however, laborious and probably result in an incomplete library.

Oligonucleotide mutagenesis of heteroduplex or cassette type (4, 5, 6, 7), although in principle efficient, is both very labour intensive and expensive, particularly when complete mutant libraries or whole genes are required.

Recently, the technology of producing a complete library of mutants by the 'spiked oligo' method has been developed such that mutant libraries of whole enzymes can be produced (see Chapter 9). This technology, while efficient may be prohibitively expensive for many research groups or for very large proteins.

A third and more elegant method involves enzymatic misincorporation (8, 9), although the earlier methods of misincorporation mutagenesis were still not efficient enough to permit complex screening systems for randomly mutagenized genes and their protein products.

This chapter describes an efficient and economical random enzymatic mutagenesis method to saturate specifically a cloned gene or defined region of DNA with random point mutations in a controlled and efficient manner. We have used this system to study the N-terminal fragment of the *Escherichia coli* β-galactosidase gene and the α-amylase of *Bacillus stereothermophilus* and the methods described here are taken from Lehtovaara *et al.* (10) and Holm *et al.* (11). In addition, this method is being used with success by groups in both the United Kingdom and the United States.

2. Principle of the method

The principle of the random misincorporation mutagenesis is illustrated in *Figure 1*. The gene to be mutagenized is in a single-stranded DNA vector and an oligonucleotide primer is hybridized downstream of the target region. The first step of the method base-specifically generates all possible 3'-ends of the elongated primer over the desired target region. A similar approach to generate variable 3'-ends has been studied before (8, 12), but so far it has only been possible to obtain some mutations localized in very short target regions this way. We carry out the enzymatic elongation of the primer in four separate sets of reactions, each of which generates a population of molecules terminating just before a given type of base, due to limiting amounts of that nucleotide. The four molecular populations are analogous to those synthesized in dideoxy sequencing, except that the next base to be incorporated is known and the 3'-termini can be further elongated in the following misincorporation reaction. In fact, the limited elongation reaction is related to the first approach to DNA sequencing, the 'plus and minus method' of Sanger (13).

In the second step point mutations are introduced to the four molecular populations by enzymatic misincorporation of the three wrong nucleotides under conditions where proof-reading does not occur. The third step involves completion of the mutagenized molecules to forms that can then be selected and amplified by molecular cloning. Selection of the mutant strand is achieved by the uracil template method of Kunkel (14) and which is described in Chapter 2.

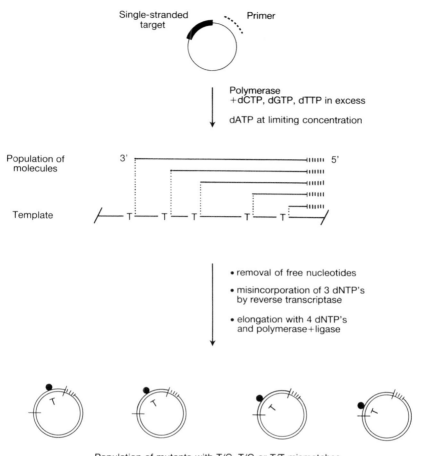

Figure 1. Principle of the method demonstrated for mutagenesis of template T nucleotides. A similar strategy is used to mutagenize template A, C, or G nucleotides in a separate series of reactions. The bases to be mutagenized in the A⁻ reaction are determined by the limited elongation step in which the primer is elongated with polymerase at various limited dATP concentrations so that the elongation will stop at different positions where an A would be required. After removal of free nucleotides, enzymatic misincorporation to these 3′-ends is carried out using three wrong nucleotides. The molecules are finally converted to covalently closed double-stranded forms which are cloned to obtain the mutant library.

2.1 Advantages of the method

This random mutagenesis method has the following advantages compared to other current methods. The regeneration of wild-type can be efficiently avoided by excluding the correct base from the misincorporation. Also, precise borders can be set to the desired target by choosing suitable molecular populations from the limited elongation. The target size can be easily varied. Moreover, the vector is not mutagenized.

The α-fragment of *E. coli* β-galactosidase was used as the target in the development of the mutagenesis procedure since it provides a simple screenable system. The region coding for the α-fragment is included in the DNA of the single-stranded phage vector M13mp19 (15, 16). The activity can be directly screened for light blue and white mutant phenotypes on indicator plates. Blue plaques indicate wild-type α-fragment sequences, or phenotypically silent mutations. In order to characterize the precise yield, distribution and equilibrium of the various mutations obtained a large number of phage DNAs from the white, light blue, and blue plaques from various mutagenesis experiments were isolated and sequenced.

3. Practical aspects of the method

3.1 Isolation of template DNA

The method for preparing uracil-containing DNA detailed in *Protocol 1* is slightly modified from that of Kunkel (14).

Protocol 1. Preparation of the single-stranded uracil-containing template

1. Dilute 5 ml of an overnight culture of the *E. coli* strain BW313 (*dut, ung, thi*-i, *relA, spoT*1/F′*lysA*) to 200 ml with $2 \times TY^a$ and grow for 30 min at 37°C with vigorous shaking.

2. Add uridine to 0.25 μg/ml and infect with the single-stranded phage stock, prepared according to Kunkel (14; see Chapter 2), at a m.o.i. of 5.

3. Continue growth at 37°C with vigorous shaking for 6 h, then pellet the cells (e.g. 5000 r.p.m., Sorvall GSA rotor, 15 min, 4°C).

4. Test the supernatant for titre in $dut^- ung^-$ (e.g. BW313) and $dut^+ ung^+$ (e.g. JM109) strains (see also Chapter 2).

5. Perform a second cycle of phage growth by repeating steps 1 to 4 but using the phage stock from step 4 of the first cycle for infection.

6. Make the supernatant containing the phage 0.5 M in NaCl and 8% (w/v) in PEG 6000, keep at 0°C for 1 h and centrifuge (e.g. 10 000 r.p.m., GSA rotor, 15 min, 4°C).

7. Dissolve the phage pellet in 30 ml TE (10 mM Tris–HCl (pH 8.0), 1 mM Na_2 EDTA) and add NaCl to 0.5 M and PEG to 2% (w/v).

8. After 1 h at 0°C collect the phage by centrifugation (e.g. 15 000 r.p.m., Sorvall SS-34 rotor, 10 min, 4°C), resuspend in 3 ml of 10 mM Tris–HCl

(pH 8.0), 50 mM NaCl, 1 mM EDTA and transfer to a 14 ml polypropylene tube.

9. Extract the solution twice with an equal volume of phenol, then twice with phenol/chloroform/isoamyl alcohol (25:24:1) and finally three times with ether.

10. Ethanol precipitate the DNA and wash with 70% ethanol, dissolve in TE and store at −20 or −70°C.

a 2 × TY: 16 g bacto-tryptone, 10 g yeast extract, 5 g NaCl per litre.

It is usually more efficient to prepare large amounts of DNA at one time since each batch can have slightly different behaviour during the elongation step (*Protocol 2*).

3.2 Base-specific limited elongation

It is important that the template and primer are very pure and no inhibitors of the polymerase are present in the reaction mixture. When the only limiting factor is the concentration of one of the four nucleotides the elongation reaction is very reproducible and can be precisely controlled.

Protocol 2. Base-specific limited elongation

1. Phenol extract and ethanol precipitate the synthetic oligonucleotide primer (17–20 nucleotides long, 2 μg).

2. Phosphorylate the 5'-end with kinase according to (18) and include approx 0.5 μCi [γ-^{32}P] ATP to label it to low specific activity.

3. Purify the primer by gel filtration in a Biogel P-6 column (1 × 20 cm) in TE according to (18) and ethanol precipitate (18).

4. Anneal the uracil-containing single-stranded template (2.5 pmol) and the primer (5 pmol) in 25 μl 15 mM Tris–HCl (pH 8.0), 7.5 mM MgCl$_2$ by heating for 10 min at 80°C and slowly cooling to ~30°C.

5. Add 130 μl water and 25 μl 100 mM Tris–HCl (pH 8.0), 50 mM MgCl$_2$, and divide the sample to 12 tubes of 15 μl (0.2 pmol) aliquots. This is sufficient to produce three different molecular populations for each of the A⁻, C⁻, G⁻, and T⁻ reactions.

6. For the A⁻ reaction, add 3 μl of a mixture containing dCTP, dGTP and dTTP or the corresponding thionucleotides each at 0.33 mM concentration to three tubes.

7. Add 2 μl of 0.75 μM dATP^a (30 Ci/mmol [α-^{32}P]dATP or 300 Ci/mmol [α-^{35}S]dATP) to the first tube, 2 μl of 2.0 μM dATP^a to the second, and 2 μl of 3.25 μM dATP^a to the third tube.

Protocol 2. continued

8. Add $0.4\,\mu$l of Klenow polymerase (2 U, Boehringer) and incubate the samples for 5 min at room temperature.

9. Stop the reactions by adding $1.2\,\mu$l of 0.3 M EDTA.

10. The C^-, G^-, and T^- reactions are performed similarly but changing the limiting nucleotide.

[a] Concentrations other than the above (final concentrations of limiting nucleotide: 0.075, 0.2, and $0.325\,\mu$M) can also be tested to find the optimal conditions for the particular sequence. (We have obtained good random molecular populations ranging to ~250 nucleotides from the 3'-end of the primer by combining one reaction made at $0.075\,\mu$M limiting concentration and two reactions made at $0.325\,\mu$M). Since each limiting nucleotide contains some label, aliquots of the reactions can be rapidly analysed on 6% sequencing gels (17) for the precise size distribution of each population (see *Figure 2*). Suitable individual A^- reactions are then combined on the basis of the desired target region. Suitable C^-, G^-, and T^- reactions are also combined respectively.

The limited elongation can also be carried out using the modified T_7 DNA polymerase (Sequenase™, United States Biochemical Corporation), but in our experience Klenow polymerase gives better results.

Figure 2 shows typical results from limited elongation where three or four different concentrations of the limiting nucleotide have been tested. The optimal concentrations, which depend on the size of the desired target and to some extent on its sequence, can easily be found empirically. The pattern of bands resembles that obtained from dideoxy sequencing of the same template but the fragments are one nucleotide shorter as expected (*Figure 2*, panel A). Limited elongation with thionucleotides gives identical molecular populations. In a system containing well-purified components the background from non-specific termination remains very low and usually there are no major differences in band intensities (*Figure 2*, panel B).

Figure 2. Analysis of base-specifically terminated reactions. The products from the limited elongation step have been electrophoresed in 6% sequencing gels. The sizes of the fragments are given as the number of nucleotides from the 3'-end of the primer. (A) The single-stranded uracil-containing M13mp19 template has been elongated with Klenow polymerase using the primer GAAGGGCGATCGCTTGCG as described in *Protocol 2*. The limiting nucleotide, [α-^{32}P]dCTP (diluted with the non-radioactive nucleotide to 30 Ci/mmol) was used at three different concentrations: 0.075, 0.2, and $0.325\,\mu$M (lanes 1–3 respectively). Aliquots of the samples were analysed on sequencing gel. In lanes 4–6 the limiting nucleotide was [α-^{32}P]dGTP (30 Ci/mmol). The marker lanes C and G show the corresponding ^{32}S-labelled dideoxy sequencing reactions respectively. (B) Template DNA, here containing an insert coding for α-amylase from *B. stereothermophilus*, has been copied from two different 20-mer oligonucleotide primers. The limiting nucleotide [α-^{35}S]dATP (diluted to ~300 Ci/mmol) was used at four different concentrations of 0.03, 0.075, 0.20, and $0.32\,\mu$M in lanes 1–4 from one primer, and in lanes 5–8 from a second primer respectively).

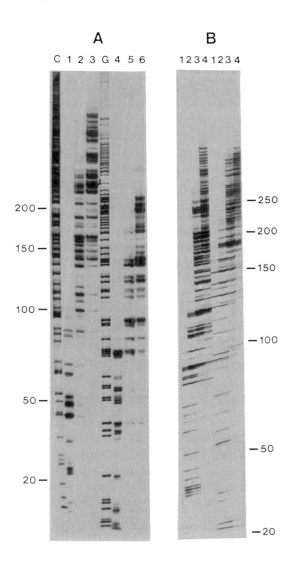

The window of the target sequence to be mutagenized can be regulated from only a few nucleotides to a few hundreds of nucleotides, or even much longer if required. This is achieved by varying the concentration of the limiting nucleotide and combining the suitable products after size analysis.

It is possible to cover a complete gene 2 kbp long with extension products from only one primer but it is often more practical to use primers at 250- to 400-nucleotide intervals in separate reactions. This enables setting of more precise borders to the target sequence. Moreover, single-track sequencing of mutant DNAs derived from a particular primer will rapidly give the precise information about the localization and frequency of mutations. In addition, these primers can be used to identify particular mutants of interest if sub-libraries generated by each oligo are kept separate during screening.

3.3 Misincorporation and elongation

The four sample fractions from limited elongation, A⁻, C⁻, G⁻, and T⁻ respectively (*Protocol 2*), are separately subjected to misincorporation. The misincorporation reaction (*Protocol 3*) of the A⁻ sample includes dCTP, dGTP, and dTTP, or the corresponding thionucleotides. We have so far mainly used reverse transcriptase, which lacks the proof-reading $3' \rightarrow 5'$ exonuclease activity and which, once incorporated, cannot remove a non-complementary nucleotide. Alternatively, thionucleotides and a polymerase unable to proof-read can be used. We have obtained better mutant yields when three nucleotides have been used in the misincorporation, rather than only one.

The correct nucleotide is omitted from the misincorporation reaction to prevent regeneration of wild type while increasing the mutant yield. It is also possible to alter the ratio of the three non-complementary nucleotides to optimize the ratio of different base substitutions. Mismatched T/G pairs, stabilized by a hydrogen bond are formed especially easily in the mutagenesis of template T's and G's. *Table 1* shows that if the three non-complementary nucleotides are present in equimolar amounts such mismatches are formed almost exclusively and also several successive mutations are observed. To randomize the mutations the concentration of the misincorporating dTTP and dGTP has to be reduced considerably, relative to the concentrations of the two other misincorporating nucleotides.

The mutation C to G, which requires the formation of a C/C mismatch, is probably difficult to obtain since this was seen at a much lower frequency than the others. However, the G to C mutations from the other strand can be obtained without difficulty. With this possible exception it seems that the remaining 11 different types of base-substitution mutants can be obtained from optimized mutagenesis of one strand only.

Table 1. Effect of the nucleotide concentrations at the misincorporation step on the equilibrium between the various mutants

Template residue to be mutagenized	Concentrations of misincorporating nucleotides	Yield of mutants	Average no. of mutations per mutant	Template mutations found	
T	dCTP 100 μM	90%	3.0	T–G	1
(A reaction)	dGTP 100 μM			C	33
	dTTP 100 μM			A	1
	dCTP 100 μM	42%	1.4	T–G	9
	dGTP 0.2 μM			C	30
	dTTP 100 μM			A	11
A	dATP 100 μM	27%	1.1	A–T	45
(T reaction)	dCTP 100 μM			G	41
	dGTP 100 μM			C	19
C	dATP 100 μM	59%	1.7	C–T	20
(G reaction)	dCTP 100 μM			G	0
	dTTP 100 μM			A	12
	dATP 20 μM	ND[a]	1.8[b]	C–T	12
	dCTP 150 μM			G	1
	dTTP 50 μM			A	21
G	dATP 100 μM	74%	2.6	G–T	0
(C reaction)	dGTP 100 μM			C	1
	dTTP 100 μM			A	58
	dATP 150 μM	9%	1.2	G–T	5
	dGTP 100 μM			C	26
	dTTP 0.2 μM			A	8

All the results are obtained from the sequencing data. These results were collected during the development of the method using the *lacZ* fragment sequence as the target. The conditions used at the misincorporation step are given in *Protocol 2*.

[a] ND = not determined
[b] The number represents the white, mutant phenotype only. Blue phenotypes were not sequenced.

Protocol 3. Nucleotide misincorporation and elongation

1. Phenol extract the four samples (A⁻, C⁻, G⁻, and T⁻) from the limiting elongation reactions (*Protocol 2*) and precipitate three times with ethanol from 250 μl of 2 M ammonium acetate using glycogen carrier if necessary. This step efficiently removes free nucleotides.

2. Adjust the samples to 0.5 pmol/25 μl in reverse transcriptase buffer (18) containing 50 mM Tris–HCl (pH 8.3), 6 mM $MgCl_2$, 60 mM KCl, 1 mM dithiothreitol, 90 μg/ml bovine serum albumin and the three misincorporating nucleotides.[a]

3. Add 10 units of reverse transcriptase (Boehringer) and allow the misincorporation to proceed at 42°C for 1.5 h, add a fresh 10 U aliquot of reverse

Protocol 3. *continued*

transcriptase and incubate for a further 1.5 h. Alternatively, thionucleo-tides can be misincorporated by Klenow polymerase at 37 °C.

4. Add all four dNTPs to 0.5 mM and incubate for at least 20 min at 37 °C to perform the chase.

5. Dilute the samples with 1 vol. of reverse transcriptase buffer (step 2).

6. Add ATP to 0.5 mM, T4 DNA ligase (Boehringer) to 200 U/ml and Klenow polymerase (Boehringer) to 140 U/ml.[b]

7. Incubate overnight at 14 °C to allow elongation to completely double-stranded molecules.

[a] We currently use the following concentrations of misincorporating nucleotides. **A⁻ reaction:** 50–100 μM dCTP, 0.4 μM dGTP and 200 μm dTTP. **C⁻ reaction:** 200 μm dATP, 50 μm dGTP and 1 μm dTTP. **G⁻ reaction:** 50 μM dATP, 250 μM dCTP and 100 μM dTTP. **T⁻ reaction:** 50 μM dATP, 100 μM dCTP and 200 μM dGTP.
[b] Modified T₇ DNA polymerase (United States Biochemical Corporation) can also be used for the final elongation.

3.4 Transformation

Competent *E. coli* JM 109 cells prepared according to Hanahan (19) are transformed with the elongation reaction from *Protocol 3*. We usually trans-form with the total reaction mix however a smaller aliquot can be used and the remaining reaction mix stored at −20 °C for subsequent transformations. The transformation frequencies obtained after the *in vitro* mutagenesis steps should be relatively high. In our experiments the double-stranded control M13mp19 or M13mp18 yielded 10^7 phage transformants/μg DNA, the back-ground from single stranded uracil template was only 0–60/μg, and the frequency obtained for the mutagenized DNA varied between 10^4 and 10^6 transformants/μg DNA.

3.5 Mutant characterization

Characterize random mutants by sequencing only; for example, A⁻ reaction products can be analysed by single track A sequencing using the same primer which was used for mutagenesis. Mutants can be identified easily since they are missing one or two bands from the wild-type pattern.

All the 12 possible types of point mutations should be obtained:

$$
\begin{array}{cccc}
 & C & & A \\
A \Rightarrow & G & C \Rightarrow & G \\
 & T & & T \\
\\
 & A & & A \\
G \Rightarrow & C & T \Rightarrow & C \\
 & T & & G \\
\end{array}
$$

The equilibrium between the different point mutations can be altered by changing the concentrations of 'wrong' nucleotides in the misincorporation reaction as shown in *Table 1*. Deletions of one nucleotide were noticed especially when the template contained a row of T nucleotides, but the frequency of such deletions was low (~1% of the mutants).

Forcing the misincorporation reaction towards completion increases the proportion of multiple mutants (*Figure 3*), while an incomplete reaction

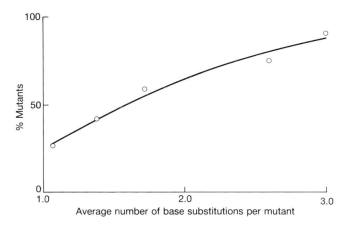

Figure 3. Correlation found between the total yield of mutant clones and the number of base substitutions in an average mutant. The values were obtained from the sequencing data. The number of mutations per template can be regulated by adjusting the nucleotide concentrations of the misincorporation step as shown in *Table 1*.

increases the wild-type background. 40–60% yield of mutants is a good compromise which gives 1.4–1.8 mutations/template. When the aim is to obtain single amino acid changes in the corresponding protein these values are close to optimal since one-quarter of the basic substitutions in codons are silent.

The method is thus highly efficient and gives both large numbers of transformants and good mutant yields, as shown in *Figure 4*, with the minimum number of manipulations.

4. Conclusion

The random *in vitro* mutagenesis method described in this chapter is well-suited for the efficient economical generation of mutant gene libraries. It requires only small amounts of DNA and few manipulations. It is completely enzymatic and does not involve the use of chemical mutagens. It is possible to (a) distribute the base substitutions randomly, (b) avoid copying the wild-type

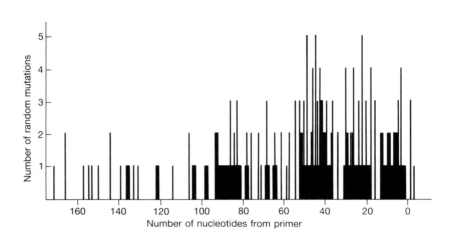

Figure 4. *In vitro* mutagenesis of the *lacZ* α-peptide. The distribution of point mutations in randomly picked mutants is shown. The primary target covered approximately 100 nucleotides from the primer. In some cases mutagenesis was allowed to proceed further upstream.

sequence in the misincorporation reaction, (c) optimize the balance between the various base substitutions, and (d) ensure that only the target region of interest is mutagenized. In addition, sequences in the vector are not mutagenized. Until now random mutagenesis has mainly been used in special cases where easy screening methods are available. The efficiency of this method is such that screening of a complete library of single base mutants can be carried out by microtitre plate assays.

References

1. Shortle, D. and Nathans, D. (1978). *Proceedings of the National Academy of Sciences of the USA*, **75**, 2170.
2. Kadonaga, J. T. and Knowles, J. R. (1985). *Nucleic Acids Research*, **13**, 1733.
3. Myers, R. M., Lerman, L. S., and Maniatis, T. (1985). *Science*, **229**, 242.
4. Botstein, D. and Shortle, D. (1985). *Science*, **229**, 1193.
5. Smith, M. (1985). *Annual Reviews in Genetics*, **19**, 423.
6. Wells, J. A., Vasser, M., and Powers, D. B. (1985). *Gene*, **34**, 315.
7. Hutchison, C. A., Nordeen, S. K., Vogt, K., and Edgell, M. H. (1986). *Proceedings of the National Academy of Sciences of the USA*, **83**, 710.
8. Zakour, R. A. and Loeb, L. A. (1982). *Nature*, **295**, 708.

9. Shortle, D. and Lin, B. (1985). *Genetics*, **110,** 539.
10. Lehtovaara, P., Koivula, A., Bamford, J., and Knowles, J. (1988). *Protein Engineering*, **2,** 63.
11. Holm, L., Koivula, A. K., Lehtovaara, P. M., Hemminki, A., and Knowles, J. K. C. (1990) *Protein Engineering*, **3,** 181.
12. Skinner, J. A. and Eperon, I. C. (1986), *Nucleic Acids Research*, **14,** 6945.
13. Sanger, F. and Coulson, A. R. (1975). *Journal of Molecular Biology*, **94,** 441.
14. Kunkel, T. A. (1985). *Proceedings of the National Academy of Sciences of the USA*, **82,** 488.
15. Messing, J. (1983). In *Methods in Enzymology* (ed. R. Wu, L. Grossman, and K. Moldave), Vol. 101, pp. 20–78. Academic Press, London
16. Yanisch-Perron, C., Vieira, J., and Messing, J. (1985). *Gene*, **33,** 103.
17. Sanger, F., Nicklen, S., and Coulson, A. R. (1977). *Proceedings of the National Academy of Sciences of the USA*, **74,** 5463.
18. Sambrook, J., Fritsch, E. F., and Maniatis, T. (ed.) (1989). *Molecular Cloning: A Laboratory Manual* (2nd edn). Cold Spring Harbor Laboratory, Cold Spring Harbor, New York.
19. Hanahan, D. (1985). In *DNA Cloning—A Practical Approach* (ed. D. M. Glover), Vol. 1, pp. 109–36. IRL Press, Oxford.

9

Spiked oligonucleotide mutagenesis

STEPHEN C. BLACKLOW and JEREMY R. KNOWLES

1. Introduction

1.1 Overview and theory

Two approaches exist for the mutagenesis of cloned genes. In site-specific mutagenesis the presumed importance of a single amino acid is probed by changing or deleting it and examining the functional consequences. This approach is often helpful in providing experimental evidence for mechanistic suggestions developed from structural and kinetic studies. Random mutagenesis, on the other hand, does not rely on any preconceived notions based on structure or mechanism, but rather is employed to identify all possible changes that satisfy a particular functional selection. These two approaches are therefore complementary and both have been used in this laboratory for studies of structure:function relationships in the glycolytic enzyme triosephosphate isomerase.

In this chapter we describe a method for the random mutagenesis of any gene that has been cloned and expressed in *Escherichia coli*. Several different methods have been devised for the random mutagenesis of cloned genes (1). These include *in vivo* methods such as the use of mutator strains, *in vitro* enzyme-dependent approaches (e.g. ref. 2, and Chapter 8), and procedures that use chemicals to damage the DNA (e.g. ref. 3, and Chapter 7). However, it is known that mutagenesis by these procedures may result in an uneven distribution of mutations because of the presence of mutagenic 'hot spots' in the DNA. The method described in this chapter was developed to ensure complete randomness in the changes of nucleotides in the gene of interest.

The approach that we have developed uses 'spiked' oligonucleotide primers to generate a library of mutants (4). A flow chart that outlines the method is given in *Figure 1*. For this procedure the target gene must lie on a single-stranded template. Rescue of single-stranded DNA is possible from several systems; the phagemid pBS(+) (*Figure 2*), and derivatives of M13 phage (*Figure 3*) are two potential vectors. Long oligonucleotides are synthesized by a protocol that gives a defined probability of incorporating one of the three 'wrong' nucleotide bases **at each position** along the oligonucleotide. These oligonucleotides are then used as primers in an efficient protocol for oligonucleotide-

177

Spiked oligonucleotide mutagenesis

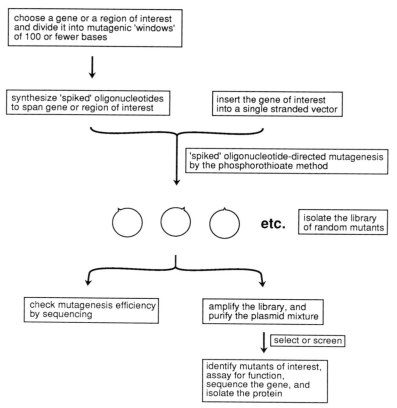

Figure 1. Flow chart for spiked oligonucleotide mutagenesis. Each step of the procedure is discussed in the text.

directed mutagenesis using a single-stranded template (ref. 5, and Chapter 3). A collection of mutants is obtained that contains changes spanning the whole length of the primer. Two major advantages of the spiked primer approach are that:

(a) the method is highly efficient in that most of the clones in the library are mutant; and

(b) the degree to which the library contains truly random mutants is easily ascertained.

It is simply necessary to sequence a number (10 to 20) of *un*selected clones to be certain that the mutagenesis procedure has given a population of mutants that is random with respect to position (within the oligonucleotide 'window'), with respect to the replaced base, and with respect to the newly-introduced base. A deliberately mismatched silent 'marker' base can also be included in the oligonucleotide in order to allow the investigator to establish, at an early

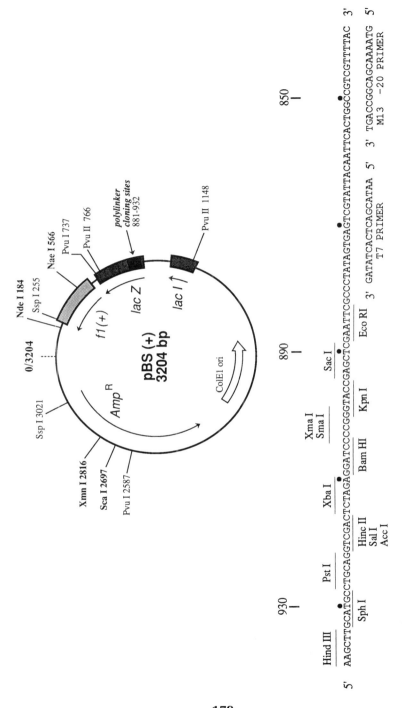

Figure 2. Diagram of the pBS(+) vector (from Stratagene). Unique restriction sites are indicated using bold type. The sequence of the polylinker region is detailed below the plasmid map. Only the sequence of the rescued strand is shown. All restriction sites in the polylinker region are unique. Arrows indicate the direction of transcription of each gene.

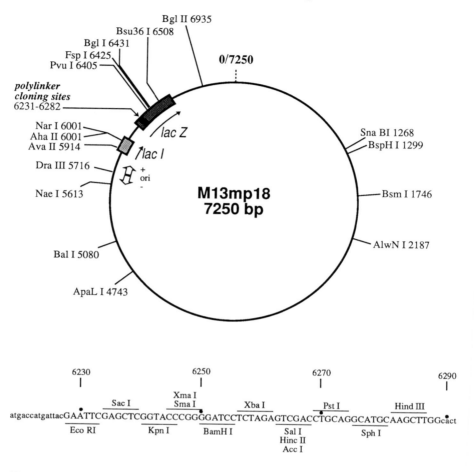

Figure 3. Restriction map of the M13mp18 vector (6, 7). Only unique restriction sites for which commercially available enzymes exist are shown. The sequence of the multiple cloning site, beginning with the initiation codon for the *lacZ* gene, is given below the plasmid map.

stage, the efficiency of the mutagenesis procedure. An overview of the protocol of Eckstein (5) for phosphorothioate mutagenesis using spiked oligo-nucleotides is given in *Figure 4*. By using several primers the whole of a gene of interest can be subjected to mutagenic variation. In this way the functional importance of different parts of the sequence may be identified.

To generate random mutations across the whole of a structural gene we divide the coding region into primer 'windows' of about 75 nucleotides and synthesize oligonucleotides of this length that are then used as primers in the phosphorothioate method of directed mutagenesis (see Chapter 3). The

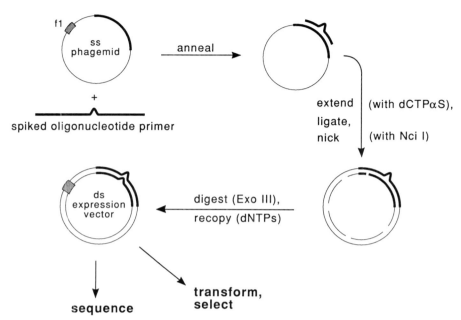

Figure 4. Schematic illustration of random mutagenesis with spiked oligonucleotides and a phagemid template in the phosphorothioate method of site-directed mutagenesis developed by Taylor *et al.* (5).

probability P of finding n errors in an m-long oligonucleotide that is synthesized with a fraction α of the three wrong nucleotides at each position is:

$$P(n,m,\alpha) = [m!/(m-n)!n!][\alpha]^n[1-\alpha]^{m-n} \qquad (1)$$

Specifically, if mutagenic 75-base primers are produced with a total contamination level of 2.5% of the three wrong phosphoramidites (i.e. 0.83% of each of the three wrong phosphoramidites), 150 000 transformants will cover >99% of the possible one-base changes and ~75% of the possible two-base changes within that oligomer window. It is always true that the number of mutagenic oligonucleotides containing one error is maximzed for α equal to $1/m$.

It is important to understand the difference between this procedure for random mutagenesis and other methods that use synthetic oligonucleotides to make a family of random mutants in several positions over a (necessarily) limited portion of the gene of interest. This 'cassette mutagenesis' approach (ref. 8, and see Chapter 10) has been used to identify the nature of the complementarity that exists between two or three nearby positions in the amino acid sequence (9). Our spiked primer approach introduces many **single** amino acid changes scattered throughout the **whole** of the gene of interest, while the cassette mutagenesis procedure produces single and multiple mutants in a relatively much smaller region of the gene. The cassette approach is most useful for examining the effect of each of the 20 amino acids at a single site, and for investigating the nature of pairwise interactions between positions in the amino acid sequence.

1.2 Strategy for spiked oligonucleotide-directed mutagenesis

The gene to be mutagenized must first be subcloned into a suitable single-stranded vector. It is most desirable to put the gene into a vector that can also be used for expression. For this purpose we chose the pBS(+) phagemid (supplied by Stratagene), from which single-stranded DNA can be rescued specifically by co-infection of pBS(+)-transformed cells with a helper phage. For example, we engineered a plasmid, derived from pBS(+), that carries the triosephosphate isomerase gene behind a *trc* promoter (see *Figure 5*). We recommend using a phagemid both for mutagenesis and for expression if possible, because tiresome and inefficient digestion and ligation steps can thus be avoided.

After preparation of single-stranded DNA (Section 3.2), the oligonucleotide-directed mutagenesis procedure (Section 3.3) is performed with the spiked oligonucleotide primer (Section 3.1). Several transformants (that have **not** been subjected to any functional selection) are first sequenced (Section 3.5) in order to determine the efficiency of the mutagenesis protocol, and to define the degree of randomness of the changes produced. The double-stranded DNA produced by the mutagenesis procedure is then amplified in a suitable strain of wild-type *E. coli* (Section 3.6). After isolation and purification (Section 3.7) the mixture of double-stranded mutant plasmids is used to transform the strain in which the selection step is done.

If a phagemid is not used as the mutagenesis and expression system then the region being subjected to mutagenesis should be subcloned into an M13 vector (6). Spiked oligonucleotide-directed mutagenesis can then be performed on the gene in M13, and the selection step performed after transfer of the gene into a suitable expression system.

2. Materials

2.1 Buffers and solutions

(a) TE; 10 mM Tris–base, 1 mM EDTA (disodium salt). Adjust pH to 8.0 with 1 M HCl.

(b) 5 × TBE; Tris–base (54 g), boric acid (27.5 g), and 20 ml of 0.5 M EDTA (disodium salt), adjusted to pH 8.0 with 5 M NaOH. Dissolve in a total volume of 1 litre of distilled water.

(c) Solution I; Glucose (50 mM), Tris–base (25 mM), and EDTA (10 mM), in distilled water. Adjust pH to 8.0 with HCl (2 M).

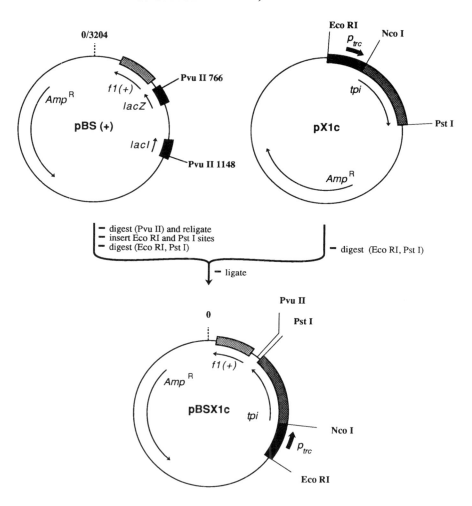

Figure 5. Diagram illustrating the construction of a phagemid expression vector for the chicken triosephosphate isomerase gene. The multiple cloning site was removed from the pBS(+) vector on a *Pvu*II fragment. Following site directed mutagenesis to introduce unique *Eco*RI and *Pst*I sites near the *Pvu*II site, the triosephosphate isomerase gene, expressed from a *trc* promoter, was introduced into the phagemid on an *Eco*RI to *Pst*I fragment.

(d) Solution II; 0.2 M NaOH containing SDS (1%, w/v). Prepare immediately before use by adding 1 ml of an aqueous solution of SDS (10%, w/v) to 9 ml of an aqueous solution of NaOH (0.22 M).

(e) Solution III; 5 M KOAc, pH 5.6. Prepare by mixing 294.5 g of KOAc and 114.5 ml of glacial acetic acid with distilled water to give a total volume of 1.0 litre Store at 4°C.

(f) 10 × kinase buffer; 1 M Tris–base adjusted to pH 8.0 with 6 M HCl, containing $MgCl_2$ (100 mM), and dithiothreitol (70 mM).

(g) 2 × oligo loading buffer; 1 × TBE, containing urea (7 M) and Bromophenol Blue (0.02%, w/v).

(h) Alkaline phenol; prepare as described by Maniatis *et al.* (10).

(i) TFB solution (11). Use the best grades of reagents commercially available. Suggested suppliers are given in parentheses.

10 mM K-MES (2[N-morpholino]ethanesulphonic acid, Sigma Chemical Co.), containing ultrapure KCl (100 mM; Aldrich Chemical), $MnCl_2 \cdot 4H_2O$ (45 mM; Aldrich), $CaCl_2 \cdot 2H_2O$ (10 mM; Fisher Scientific), and hexamine cobalt (III) trichloride ($HACoCl_3$, 3 mM; Aldrich).

To prepare the solution, combine 7.4 g KCl, 8.9 g $MnCl_2 \cdot 4H_2O$, 1.5 g $CaCl_2 \cdot 2H_2O$, and 0.8 g hexamine cobalt (III) trichloride with 20 ml of a 0.5 M solution of MES, adjusted to pH 6.3 using solid KOH, and add glass distilled water to give a total volume of 1 litre. Stir to dissolve, and sterilize by filtration through a 0.22 μ membrane. Store at 4°C.

(j) DnD solution (11). Suggested suppliers are listed in parentheses.

10 mM potassium acetate (Aldrich), pH 7.5, containing dithiothreitol (1.0 M, Calbiochem), dissolved in dimethyl sulphoxide (Fluka Chemical Co). To prepare the solution, combine 100 μl of a 1.0 M stock (pH adjusted to 7.5 using solid KOH) of potassium acetate with 1.53 g dithiothreitol, and bring the volume to 10 ml with dimethyl sulphoxide.

2.2 Media

(a) YT; Dissolve 8 g bacto-tryptone, 5 g bacto-yeast extract, and 5 g of NaCl in 1 litre of distilled water. Sterilize by autoclaving. For plates add 15 g agar per litre of YT broth. For top agar add 0.6 g agar per 100 ml of broth. Add antibiotics by sterile filtration as needed.

(b) SOB; Dissolve 20 g bacto-tryptone, 5 g bacto-yeast extract, 60 mg NaCl, and 190 mg KCl in 1 litre of glass distilled water. Sterilize by autoclaving.

Prepare a 2 M stock of Mg^{2+}, as an aqueous solution containing $MgCl_2$ (1 M) and $MgSO_4$ (1 M). Sterilize by filtration through a 0.22 μ membrane. Store at −20°C. Add 1 ml of this solution to the medium just before use.

(c) SOC; Add glucose to a final concentration of 20 mM (from a 2 M stock aqueous solution, sterilized by filtration through a 0.22 μ membrane and stored at −20°C) to SOB medium (containing Mg^{2+}) immediately before use.

2.3 Bacterial strains

(a) *Escherichia coli* strain DH5 [*end*A1, *rec*A1, *hsd*R17 (r_k^-, m_k^+), *supE44*, *thi*-1, λ^-, *gyrA96*, *relA1*, F^-] (12)

(b) *E. coli* strain TG1 [K12, Δ(*lac-pro*), *sup*E, *thi*, *hsd*D5/F′ *tra*D36, *pro*A^+B^+, *lac*I^q, *lacZ*ΔM15] (13).

3. Procedures

3.1 Preparation of oligonucleotides for mutagenesis

It is necessary to have access to an automated DNA synthesizer to make the spiked oligonucleotides to be used in the mutagenesis procedure. [For a review of automated DNA synthesis using phosphoramidite chemistry, see Caruthers (14).] The automated synthesis proceeds from the 3'-end; consequently, for oligonucleotides greater than 25 bases in length, it is important to remember to remove the final trityl protecting group from the 5'-end of the oligonucleotide after completion of the final coupling cycle. For oligonucleotides with fewer than 25 bases, purification and detritylation can be achieved in one step using the oligonucleotide purification cartridges supplied by Applied Biosystems Inc.

3.1.1 Spiked oligonucleotide synthesis

In this section, we describe the procedure for preparing oligonucleotides with a defined probability of containing one of the three 'wrong' bases at a given position. The procedure given in this example will produce an oligonucleotide with a 3% probability of having an improper base at any particular position. The oligonucleotides synthesized according to *Protocols 1* or *2* should contain 'wrong' bases incorporated randomly with respect to position, replaced base, and newly incorporated base. We recommend the inclusion in the oligonucleotide of a silent mismatching 'marker' base so that the efficiency of the oligonucleotide-directed mutagenesis protocol can be established before any time-consuming selections are embarked upon.

Before beginning synthesis of the spiked oligonucleotide prepare the DNA synthesizer for a new synthesis according to the manufacturer's instructions. We recommend the use of fresh reagents and fresh amidites when attempting the synthesis of a spiked oligonucleotide. In the synthesis of clean ('unspiked') oligonucleotides the phosphoramidites are used in excess, and it is not crucial that the chemical reactivity of all four amidites be equal. In the synthesis of a spiked oligonucleotide, however, the statistically random introduction of wrong bases relies on the competitive incorporation of an incorrect phosphoramidite into the growing chain. Thus, if the amidite reagents decompose at different rates older reagents will not give a proper distribution of wrong bases, and the fraction of wrong nucleotides introduced may not be that which is desired.

Protocol 1. Synthesis of spiked oligonucleotides on a Milligen/ Biosearch 7500 DNA synthesizer

1. Purchase several empty vials, septa, and aluminum seals (Milligen/ Biosearch cat. numbers 103036, 103030, 103031). Bake the vials dry by

Protocol 1. *continued*

heating at 110°C for several hours and allow them to cool in a desiccator under argon. Seal the vials with a crimping tool (cat. number 104009) before use. Syringes should be purged by drawing argon into the barrel and expelling it several times.

2. Use septum-sealed water-free acetonitrile (Fisher Scientific or Cruachem Inc.) to dissolve the pure amidites. Draw enough acetonitrile into a disposable 10 ml syringe (purged with argon as described in step 1), fitted with a 19-gauge needle, to dissolve the pure amidites to a final concentration of 1 g/10 ml.

3. To prepare a vial of mixed amidites providing a contamination level of 3% at each position withdraw 300 μl of each of the four pure amidite reagents and add it to an empty, dry vial, prepared as in step 1. Bring the total volume of liquid in the vial to 10 ml with septum-sealed acetonitrile. The total concentration of all four amidites in the 'mix' vial, relative to the concentration of dissolved amidite in each of the 'pure' vials, should be roughly equal to four times the desired probability of incorporating a 'wrong' nucleotide at any position. Thus, to make mixtures that give contamination levels other than 3% adjust the amount of each pure amidite added to the mix vial.

4. Pack a glass column (Milligen/Biosearch cat. number 101936), fitted with a frit assembly at one end, with bulk controlled-pore glass resin with the desired 3'-base attached. For oligonucleotides of more than 35 bases use low-loading resin (0.2 μmol scale, 500 Å pore-size; cat. number 901001). Leave a space of about 3 mm at the top of the column. Screw the other frit assembly to the top of the column and place the column on the path for synthesis on column no. 1. Place an empty column in line on the path to column no. 2.

5. Check to make sure that the columns do not leak by sending acetonitrile to column no. 1 for 30 sec.

6. Run the synthesis of the 'spiked' oligonucleotide on column no. 1. Include a detritylation step after the last cycle for oligonucleotides longer than 25 nucleotides.

We have written a computer program that directs automated synthesis of spiked oligonucleotides on a Milligen/Biosearch 7500 DNA synthesizer. The program is available from Milligen/Biosearch upon request. The procedure described in *Protocol 2* will give a contamination level of 3.0% on other types of DNA synthesizer.

Protocol 2. Synthesis of spiked oligonucleotides on machines other than the Milligen/Biosearch Model 7500

1. Prepare a dry vial (or septum-sealed flask) to receive the equimolar mixture of amidite reagents as in *Protocol 1*, step 1.
2. Dissolve amidite reagents at a concentration of 1.0 g/10 ml as described in *Protocol 1*, step 2.
3. By syringe, transfer 500 μl of each of the pure amidites into the empty, dry vial, and mix thoroughly.
4. Again by syringe, transfer 400 μl of this mixture back into each of the four pure reagent vials.
5. Attach these spiked vials to the DNA synthesizer.
6. Run the synthesis of the spiked oligonucleotide as for any other pure synthesis.

Note that with smaller instruments, which only have five amidite reservoirs, the ability to use pure (unspiked) reagents for specific positions in the oligonucleotide is limited. With the Milligen/Biosearch 7500 model synthesizer this problem is avoided because the pure amidite vials are separate from the mixed amidite vial and mixing only occurs in the line to the column.

3.1.2 Oligonucleotide purification

In *Protocol 3* the oligonucleotide is cleaved from the solid support and the protecting groups are removed from the nucleotide bases. The oligonucleotide may then be gel purified according to *Protocol 4*.

Protocol 3. Oligonucleotide cleavage and deprotection

1. Recover the resin containing the 'spiked' oligonucleotide into a 1.5 ml screw-cap microfuge tube.
2. Remove the acetonitrile from the screw-cap microfuge tube using a speed-vacuum concentrator (e.g. Savant Instruments Inc.)
3. Add 1.0 ml of a fresh concentrated solution of aqueous ammonium hydroxide (such as AR grade from Mallincrodt Inc.) to the resin containing the oligonucleotide in the screw-cap tube and cap the tube tightly. Seal the cap with Parafilm.
4. Incubate the tube and contents at 56°C for 12 h to remove the oligonucleotide from the solid support, and to deprotect the component bases.
5. Place the tube on ice for 10 min.
6. Pipette the supernatant into a fresh microfuge tube and use a speed-vacuum concentrator to bring the sample to 50 μl or less.

Protocol 3. *continued*

7. Dilute the sample to $50 \mu l$ with water and quantitate the amount of oligonucleotide recovered by measuring the absorbance at 260 nm of a 1/200 dilution in H_2O (a solution of 0.05 absorbance units is roughly equal to $1 \mu g/ml$). The ratio of A_{280nm}/A_{260nm} should be close to 0.5.

10% polyacrylamide gels are used to purify oligonucleotides longer than 25 bases. The gel-purified oligonucleotide is then suitable for use in the mutagenesis procedure. 20% gels can be employed to purify oligonucleotides shorter than 25 bases. A complete procedure for the preparation and running of acrylamide gels is given in Maniatis *et al.* (10).

Protocol 4. Gel purification of oligonucleotides

1. Wash 20×20 cm glass plates with soap and water. Rinse with water and ethanol (EtOH) and allow the plates to dry.

2. Set up the gel sandwich using 1.5 mm Teflon spacers.

3. Make a stock solution of 38% acrylamide, 2% bisacrylamide (w/v). Store at 4°C.

4. For a 10% gel add 36 g urea, 14.5 ml $5 \times$ TBE, 18 ml acrylamide solution and 12 ml of distilled water to a 250 ml Erlenmeyer flask. Warm the flask and swirl to dissolve the urea.

5. Filter the solution through Whatman No. 1 filter paper on a Buchner funnel. After filtration stir the solution and de-gas on an aspirator.

6. Add $300 \mu l$ of a 10% (w/v) solution of ammonium persulphate. De-gas with stirring for 30 sec.

7. Stop stirring, remove from the vacuum, and add $15 \mu l$ TEMED. Pipette the gel solution between the sandwiched plates. Fit the gel with a comb to make wells 1 cm wide by 1.5 cm deep. Allow the gel to polymerize overnight before use. (Visualization of DNA bands under ultraviolet light becomes difficult when the gel has not polymerized completely.)

8. Pre-run the gel for 30 min at 40 mA ($\sim$300 V) in $1 \times$ TBE as running buffer. Wash out the wells with running buffer using a Pasteur pipette before loading the sample on to the gel.

9. Add an equal volume of $2 \times$ oligo loading buffer to the sample from step 7 of *Protocol 3*, boil the sample for 3 min, and load up to 0.2 mg of the sample on the polyacrylamide gel.

10. Run the gel in $1 \times$ TBE at 300 V until the dark blue (Bromphenol Blue) dye is near the bottom of the gel (roughly 1.5–2 h).

11. Separate the gel plates with a screwdriver and transfer the gel onto a large piece of plastic wrap. Wrap the gel once with the plastic wrap.

12. Place the wrapped gel on top of a silica plate containing fluorescent indicator and use a short-wave UV lamp (254 nm) to identify the purified oligonucleotide (it should be the major band and lie at the highest molecular weight). Circle the area corresponding to your band with a marker pen and turn off the UV lamp.

13. Cut out the circled band carefully with a razor blade or a scalpel and discard the small pieces of plastic wrap.

14. Transfer the gel fragment to a plastic test tube (Falcon 2059 or equivalent), and crush the gel fragment with a polished glass rod until it becomes a sticky paste (about 5 min is enough).

15. Add 1 ml of distilled water to the tube and shake overnight (up to 24 h) at 37°C to leach the oligonucleotide out of the gel.

16. Transfer the sample to a 1.5 ml microcentrifuge tube and centrifuge at $\sim 12\,000\,g$ for 10 min to remove polyacrylamide particles; transfer the sample to a fresh tube and speed-vacuum concentrate the sample to a small volume ($< 250\,\mu l$).

17. Use 4 vol. of EtOH and 1/10 vol. of a 3 M solution of NaOAc (aq) to ethanol precipitate the oligonucleotide.

18. Centrifuge the sample in a microfuge for 15 min. Decant the supernatant, wash the pellet (which may not be visible) with ice-cold 100% EtOH, and drain well.

19. Dry the pellet on a speed-vacuum concentrator. Resuspend the purified oligonucleotide in $100\,\mu l$ of distilled H_2O and quantitate by A_{260nm}.

3.2 Preparation of single-stranded template (for mutagenesis)

Phagemids, such as pBS(+) (Stratagene Cloning Systems), are double-stranded plasmids from which single-stranded DNA can be rescued. The single-strand DNA is produced by coinfection with a helper phage that is defective in self-packaging; the phagemid contains a *trans*-acting element that directs packaging of its single strand. The pBS(+) phagemid has a high copy number and we have used it as a vector for both mutagenesis and expression. It is important to prepare pure single-stranded template according to *Protocol 5* to ensure high efficiency in the phosphorothioate mutagenesis procedure. Alternatively, single-stranded template DNA may be prepared from M13 clones according to *Protocol 6*.

Protocol 5. Preparation of template from phagemids

This procedure is similar to that recommended by Stratagene for the recovery of single-stranded phagemid DNA.

1. Transform an F^+ strain (e.g. TG1) with a phagemid carrying an antibiotic resistance marker, and plate on an appropriate selective plate. Incubate overnight at 37 °C.

2. Pick a colony to inoculate 5 ml of sterile rich (YT) broth containing antibiotic. Grow overnight with shaking at 37 °C.

3. Use this overnight culture to inoculate 500 ml of rich broth (not containing antibiotic). Grow until A_{550nm} equals 0.1.

4. Add VSC-M13 helper phage (Stratagene Cloning Systems) at an m.o.i. of 5 to 20 (roughly 10^{11} p.f.u.). Shake in a 37 °C incubator for 6–8 h.

5. Remove the cells from the culture by centrifugation at 8000 g for at least 20 min. Transfer the supernatant to a fresh flask taking care to avoid the cell pellet. It is not necessary to recover all of the supernatant. If the supernatant is not clear, recentrifuge to clarify.

6. Precipitate the phage by adding 0.2 vol. of an aqueous solution of 2.5 M NaCl containing 20% (w/v) PEG-6000 to the supernatant. Leave for 1 h at 4 °C.

7. Centrifuge the solution at 8000 g for 30 min to pellet the phage. Pour off the supernatant and drain the pellet well. Use a tissue or a pasteur pipette to remove traces of the supernatant. Resuspend the pellet in a total volume of about 1 ml TE, and divide equally between two microcentrifuge tubes.

8. Centrifuge for 5 min to remove cell debris. Transfer the supernatant to a fresh tube taking care not to disturb the pellet.

9. Reprecipitate the phage by adding 0.2 vol. of an aqueous solution of 2.5 M NaCl containing 20% (w/v) PEG-6000 to the supernatant. Leave on ice for 15 min. Centrifuge then discard the supernatant. Recentrifuge briefly, and use a pipette to remove traces of the supernatant. Redissolve the pellet in 300 μl of TE.

10. Add an equal volume of alkaline phenol and mix vigorously on a vortex mixer for 30 sec. Leave for 5 min and then vortex again. Centrifuge for 3 min to separate the phases. Pipette the aqueous (upper) layer into a fresh microcentrifuge tube.

11. Extract with an equal volume of chloroform/isoamyl alcohol (24:1, v/v). Vortex for 15 sec, and centrifuge briefly in a microcentrifuge to separate the phases. Transfer the aqueous layer to a new microcentrifuge tube. Remove traces of chloroform on a speed-vacuum concentrator.

12. Precipitate the single-stranded DNA by adding 3 vol. of 100% ethanol and 0.1 vol. of a 3 M aqueous solution of NaOAc. Centrifuge for 15 min to recover the DNA.

13. Wash the pellet with 100% ethanol and drain. Dry the pellet on a speed-vacuum concentrator and resuspend it in $500 \mu l$ TE.

14. Quantitate the DNA by measuring A_{260nm}. A value of A_{260nm} of 25 is equivalent to 1 mg/ml. Store the sample at 4°C.

For the large-scale production of single-stranded DNA from M13 (*Protocol 6*) it is first necessary to prepare a phage stock that can be used for subsequent infection of host cells.

Protocol 6. Preparation of single-stranded DNA from M13

A. Preparation of phage stock

1. Transfect competent TG1 with the M13 derivative as follows. Add double-stranded M13 DNA (roughly 10 ng DNA in about $10 \mu l$ of TE) to $300 \mu l$ competent cells (*Protocol 8*) in a pre-chilled, 17 × 100 mm, poly-propylene tube (Falcon 2059 or equivalent). Leave on ice for 15 to 40 min. Heat shock for 90 sec at 42°C. Chill on ice for 2 min and then add: $200 \mu l$ mid-log TG1 cells; 3 ml YT top agar (0.6%, w/v), melted and kept warm at 47°C; $20 \mu l$ IPTG (100 mM, aq.); and $50 \mu l$ of X-gal (2% (w/v) solution in dimethylformamide).

2. Gently agitate on a vortex mixer. Pour the mixture immediately on to a warm (~37°C) YT plate. Let the plate cool for 15 min and then incubate inverted at 37°C. Plaques should be visible after 8 h.

3. Grow an overnight culture of TG1. Dilute this culture 1:100 into 3 ml of fresh YT medium.

4. Remove a well-isolated plaque from the plate using a $50 \mu l$ micropipette. Dispense the plaque into the diluted culture from step 3, and incubate for 6 h with shaking (250 r.p.m.).

5. Transfer a 1.2-ml portion of the culture to a microcentrifuge tube and microfuge for 10 min. Remove the supernatant carefully into a fresh tube and discard the cell pellet. The sample from the supernatant is the phage stock and its titre should be between 10^{11} and 10^{12} p.f.u./ml.

B. Large scale preparation of template (500 ml)

6. Grow a 5 ml culture of TG1 [from a freshly streaked minimal plate (15)] overnight.

7. Add M13 phage stock (from part A, step 5) at a m.o.i. of about 20.

8. Inoculate 500 ml of sterile YT broth with the M13-infected cells.

9. Incubate this culture with shaking for 8 h at 37°C.

10. Follow steps 5 to 14 from *Protocol 5*. The total yield of single-stranded DNA should be about 0.5 mg.

3.3 Oligonucleotide-directed mutagenesis

This procedure is based on the phosphorothioate method of Taylor *et al.* (5). Instead of using an oligonucleotide designed to introduce a single mutation the spiked oligonucleotide synthesized as described in Section 3.1 is used. We have followed the protocol supplied with the site-directed mutagenesis kit from Amersham. Detailed protocols for this directed mutagenesis approach are provided in Chapters 3 and 5.

3.4 Preparation of competent cells and transformation

The two methods to prepare competent cells are those described by Hanahan (11). The $CaCl_2$ procedure (*Protocol 7*), which we use for routine transformation, is quicker and is easily adapted for the preparation of frozen competent cells. However, the efficiency of transformation is usually in the range of 10^6 to 10^7 transformants/μg of plasmid DNA. The second procedure (*Protocol 8*), which gives a transformation efficiency of $2-7 \times 10^8$ transformants/μg of plasmid DNA in the most competent strains (e.g. DH5), is used in the amplification and screening steps. *Protocol 9* describes a procedure for transformation of competent cells with DNA.

Protocol 7. Rapid preparation of competent cells with $CaCl_2$

1. Prepare a 50 mM aqueous solution of $CaCl_2$ and sterilize by filtration. Store at 4°C.

2. Streak cells of the appropriate strain from a glycerol frozen stock on to a YT plate and incubate overnight at 37°C.

3. Inoculate 5 ml of YT broth with cells from a single colony and shake overnight at 37°C.

4. Use 1 ml of this culture to inoculate 100 ml of YT broth and incubate with shaking until the culture reaches an A_{550nm} of 0.3 to 0.8 (about 1–3 h). Pour the culture into two pre-chilled 50 ml centrifuge tubes and leave these tubes on ice for 15–60 min.

5. Centrifuge the tubes at 3000 g for 10–15 min at 4°C to recover the cells. Pour off the supernatant and tap the inverted tubes on paper towels to remove as much of the supernatant as possible.

6. Resuspend the cells in ⅓ of the original volume (16 ml/50 ml of culture) of 50 mM ice-cold $CaCl_2$ solution. Store on ice for 20 min.

7. Centrifuge at 3000 g for 10 min at 4°C to recover the cells. Remove the supernatant as in step 5.

8. Resuspend the cells in 50 mM ice-cold $CaCl_2$ in 1/12.5 of the original volume (4 ml/50 ml of culture). Allow the suspension to age on ice for at least 1 h before use. Cells reach peak efficiency after 24 h and remain competent for up to a week.

9. To prepare frozen competent cells, add an aqueous solution of 80% glycerol (w/v, previously sterilized by autoclaving) to give a solution of competent cells in 15–20% glycerol. Place small (0.5 ml) microcentrifuge tubes on dry ice and add 200 μl aliquots of competent cells to each tube. Store these frozen competent cells at $-70\,^{\circ}$C and thaw on ice immediately before use.

Protocol 8. Preparation of supercompetent cells

1. Streak cells of the appropriate strain on to a rich plate and incubate at 37°C overnight.

2. Pick five colonies and disperse them all in 1 ml SOB medium. Use this suspension to inoculate 50 ml of sterile SOB medium (contained in a 500 ml flask) and incubate at 37°C with shaking until the A_{550nm} of the culture is between 0.4 and 0.8.

3. Pour the culture into a pre-chilled polypropylene centrifuge tube (50 ml). Chill on ice for 10–15 min.

4. Recover the cells by centrifugation at 750–2000g (2000–3000 r.p.m. in a clinical centrifuge) for 12–15 min at 4°C. Drain the pelleted cells thoroughly by inverting the tubes on paper towels and tapping sharply to remove as much liquid as possible.

5. Resuspend the cells in 17 ml TFB solution by mixing moderately on a vortexer. Leave on ice 10–15 min.

6. Pellet the cells and drain as in step 4.

7. Resuspend the cells in 4 ml TFB.

8. To this suspension add 140 μl of DnD solution through a pipette directed into the middle of the cell suspension and swirl for several seconds to mix. Incubate the tube on ice for 10 min.

9. Add another 140 μl portion of DnD solution as in step 8. Leave on ice 10–20 min. The cells are now ready for transformation with DNA.

Competent cells prepared according to *Protocols 7* and *8* may be transformed with DNA according to the procedure described in *Protocol 9*.

Protocol 9. Transformation of competent cells

1. Pre-chill 17 × 100 mm polypropylene tubes on ice (Falcon 2059 tubes or equivalent).

2. Add 100 μl of the competent cell suspension to each tube.

Protocol 9. *continued*

3. Pipette the sample of transforming DNA ($\sim$10 ng in 1–10 μl of TE) into the suspension of competent cells and swirl gently to mix.

4. Leave on ice for 20–40 min.

5. Heat shock the cells at 42°C for 45 sec. Place on ice for 2 min.

6. Add about 0.8 ml of sterile SOC medium to the cell suspension and allow the cells to recover at 37°C for 30–60 min, with gentle shaking.

7. Place an appropriate volume of the suspension on a selective plate. Generally, sample volumes from 50 to 150 μl give good colony densities on 15 × 100 mm plates.

3.5 Determination of mutagenesis efficiency by sequencing

We do not describe the method for dideoxy sequencing with [α-^{35}S]dATP here because it is described in detail elsewhere (17) and many suppliers provide sequencing kits that are self-explanatory. It is crucial, however, that at least 10 **unselected** clones from each oligonucleotide-directed mutagenesis reaction be sequenced in order to determine the efficiency of mutagenesis, and to establish that the predicted level of incorporation of the three wrong bases has occurred. *Protocol 10* describes the isolation of single-stranded template DNA suitable for DNA sequence analysis.

Protocol 10. Preparation of single-stranded DNA for sequencing

A. Preparation of phagemid single-strand DNA

1. Transform competent TG1 (*Protocols 7* and *9*) with the double-stranded phagemid DNA derived from the oligonucleotide-directed mutagenesis reaction (Section 3.3).

2. Inoculate 3 ml of sterile YT broth with the cells from a single colony from step 1.

3. Incubate the culture at 37°C with shaking until the A_{550nm} is between 0.1 and 0.3 (the culture should be slightly cloudy). Add $\sim$5 × 10^9 p.f.u. of VCS-M13 helper phage (Stratagene Cloning Systems) and continue to incubate the culture with shaking at 37°C for a further 6 h.

4. Transfer 1.2 ml of the culture to a 1.5 ml microcentrifuge tube and centrifuge the sample at $\sim$12 000 g for 5 min. Remove the supernatant carefully, avoiding the cell pellet, and transfer it to a fresh tube.

5. Add 240 μl of an aqueous solution of 2.5 M NaCl containing 20% (w/v) PEG-6000 to the supernatant. Mix the sample well on a vortexer. Leave for 30 min on ice.

6. Pellet the phage by centrifugation of the sample at $\sim$12 000 g for 10 min. Discard the supernatant. Microfuge the pellet for 2 min and remove additional traces of the supernatant with a pipette.

7. Redissolve the pellet in 400 μl of TE. Centrifuge the solution at $\sim$12 000 g to remove the cell debris. Transfer the supernatant to a fresh tube without disturbing the pellet.

8. Add 0.2 vol. of an aqueous solution of 2.5 M NaCl containing 20% (w/v) PEG-6000 to the supernatant, and precipitate the phage on ice for 15 min. Centrifuge for 5 min at $\sim$12 000 g to recover the phage. Discard the supernatant, centrifuge briefly in a microfuge and use a pipette to remove traces of the supernatant.

9. Resuspend the phage pellet in 200 μl of TE.

10. Add an equal volume of alkaline phenol to the phage suspension. Mix the sample vigorously on a vortexer. Centrifuge the sample at $\sim$12 000 g for 2–5 min to separate the phases. Transfer the upper (aqueous) layer to a fresh microcentrifuge tube.

11. Add an equal volume of a mixture of chloroform and isoamyl alcohol (24:1, v/v) to the sample. Vortex well. Microfuge the sample briefly to separate the phases. Transfer the aqueous layer to a fresh tube and spin this solution in a speed-vacuum concentrator to remove traces of chloroform.

12. Precipitate the single-stranded DNA by adding 3 vol. of 100% ethanol and 0.1 vol. of a 3 M aqueous solution of NaOAc to the sample from step 11. Vortex to mix and chill the suspension at -70°C for 10 min.

13. Centrifuge at $\sim$12 000 g for 15 min. Decant the supernatant and rinse the pellet with 100% ethanol. Dry the pellet in a speed-vacuum concentrator.

14. Resuspend the pellet in 20 μl of TE. A volume of 5 μl is usually enough single-stranded DNA for use in each dideoxy sequencing reaction.

B. Preparation of M13 single-strand DNA

15. Follow part A (steps 1 to 5; preparation of phage stock) of *Protocol 6*.

16. Complete the preparation of M13 single strand by following steps 5 to 14 of *Protocol 10*.

3.5.1 Preparation of the gel and gel apparatus

We have found the gel apparatus sold by Bethesda Research Laboratories (Gibco-BRL) to be convenient and easy to use. Detailed procedures for the preparation and running of DNA sequencing gels are given in *Nucleic Acid Sequencing—A Practical Approach* (17).

3.6 Amplification of mutant plasmid DNA

To amplify (and methylate) the pool of mutant plasmids from oligonucleotide-directed mutagenesis we transform supercompetent DH5 and collect plasmid DNA from the resultant colonies (according to *Protocols 11* and *12*). In this way we were able to isolate plasmid DNA which encoded approximately 1.5 million mutant proteins and could subsequently screen these mutants in a suitable host strain for selection of the desired phenotype.

Protocol 11. Preparation of amplified and methylated double-stranded DNA from mutagenesis mixtures

1. Prepare supercompetent DH5 as described in *Protocol 8* and transform according to *Protocol 9*. In this case spread the whole sample from each transformation on to large agar plates (15 × 150 mm) containing the appropriate additives to select for transformants. It may be easier to centrifuge the transformed cells in microcentrifuge tubes (for 1 min at ~7 000 g), decant the supernatant, and resuspend the cells in a smaller volume to minimize the number of platings.

2. Incubate the plates overnight at 37°C. Colonies should appear after about 12 h.

The next steps allow the recovery of the cells from the plates so that a double-stranded plasmid preparation can be made from the mixture of transformants.

3. Pour about 5 ml of sterile YT broth on to one of the plates and spread it gently with a sterilized glass spreader to suspend the cells.

4. Transfer the suspension to a sterile centrifuge tube using a sterile pipette. Repeat step 3 until the colonies from all of the plates have been recovered. At this point all suspensions resulting from directed mutagenesis using the same spiked oligonucleotide can be pooled into one tube.

5. Mix each pool of transformants thoroughly (for example in a rotary shaker for 30 min), and withdraw 3 ml from each pool for purification of the plasmid DNA (*Protocol 12*).

The following method for the isolation of supercoiled plasmid DNA is based on the procedure of Birnboim and Doly (16).

Protocol 12. Preparation of plasmid DNA by alkaline lysis

1. Pellet 3 ml of the cells from each of the plate scrapings (*Protocol 11*) in 1.5 ml microcentrifuge tubes by repeated spinning (~12 000 g for 10 sec) and decanting of supernatant. Drain the cell pellet thoroughly by inversion on to a paper towel.

2. Add $200\,\mu l$ of solution I containing $2\,mg/ml$ of freshly added lysozyme (Sigma) to each cell pellet from step 1. Vortex each sample until the cells are suspended. Leave the samples at room temperature for 10 min or longer.

3. Place the tubes on ice. Add $400\,\mu l$ of solution II to each tube. Mix by inversion and by gently swirling the tubes (do **not** vortex). Leave the tubes on ice for 1 min.

4. Add $300\,\mu l$ of solution III, pre-chilled to $4°C$, to each tube. Mix by inversion and gentle vortexing. Incubate the tubes on ice for 5 min.

5. Centrifuge the samples at $\sim12\,000\,g$ for 5 min. Transfer the supernatant ($\sim850\,\mu l$) from each tube to a fresh microcentrifuge tube. Be careful to avoid the sediment.

6. Add 0.6 vol. (about $510\,\mu l$) of isopropanol to the supernatant from step 5. Mix by inversion and vortexing. Place the samples on ice for 10 min.

7. Centrifuge at $\sim12\,000\,g$ for 5 min. Discard the supernatant and drain the pellets well. To remove traces of isopropanol subject the pellets to high vacuum for 2–5 min in a speed-vacuum concentrator.

8. Resuspend each pellet in $300\,\mu l$ of TE. Mix each vigorously on a vortexer to redissolve.

9. Add an equal volume of alkaline phenol, equilibrated with TE, to each sample. Mix vigorously on a vortexer. Centrifuge the tubes containing the samples at $\sim12\,000\,g$ for 2–5 min to separate the phases. Transfer the upper (aqueous) layers to fresh microcentrifuge tubes.

10. Add an equal volume of a mixture of chloroform and isoamyl alcohol (24:1, v/v) to each sample. Vortex well. Microcentrifuge each sample briefly to separate the phases. Transfer the aqueous layers to fresh microcentrifuge tubes and spin these samples in a speed-vacuum concentrator to remove traces of chloroform.

11. Precipitate the plasmid DNA by adding 3 vol. of 100% ethanol and 0.1 vol. of a 3 M aqueous solution of NaOAc. Vortex to mix, and chill at $-70°C$ for 10 min.

12. Centrifuge at $\sim12\,000\,g$ for 15 min. Decant the supernatant and rinse the pellet with 100% ethanol. Dry the pellet in a speed-vacuum concentrator.

13. Resuspend the pellet in $30\,\mu l$ TE. Leave at room temperature for 10 min, vortex.

This protocol should yield approximately $1–2\,\mu g$ plasmid DNA/ml of culture. This plasmid DNA is now ready for use in the functional selection.

4. Summary

This chapter describes a method for the generation of a library of random mutants in a gene or regulatory region of interest. The procedure employs automated DNA synthesis and an efficient protocol for oligonucleotide-directed mutagenesis with single-stranded vectors. This method was developed in our laboratory in the search for second-site suppressors of a sluggish mutant of triosephosphate isomerase (4).

Acknowledgements

The program for synthesis of spiked oligonucleotides on the Milligen/ Biosearch 7500 instrument was written by Dr Jeff Hermes (Present address: Merck and Co., Inc., P.O. Box 2000, Rahway, New Jersey 07065, USA), in collaboration with Milligen/Biosearch.

References

1. Botstein, D. and Shortle, D. (1985). *Science*, **229**, 1193.
2. Lehtovaara, P. M., Koivula, A. K., Bamford, J., and Knowles, J. K. C. (1988). *Protein Engineering*, **2**, 63.
3. Myers, R. M., Lerman, L. S., and Maniatis, T. (1985). *Science*, **229**, 242.
4. Hermes, J. D., Parekh, S. M., Blacklow, S. C., Köster, H., and Knowles, J. R. (1989). *Gene*, **84**, 143.
5. Taylor, J. W., Ott, J., and Eckstein, F. (1985). *Nucleic Acids Research*, **13**, 8765.
6. Messing J. (1983). In *Methods in Enzymology* (ed. R. Wu, L. Grossman, and K. Moldave), Vol. 101, pp. 20–78. Academic Press, London and New York.
7. Yanisch-Perron, C., Viera, J., and Messing, J. (1985). *Gene*, **33**, 103.
8. Wells, J. A., Vasser, M., and Powers, D. B. (1985). *Gene*, **34**, 315.
9. Reidhaar-Olson, J. and Sauer, R. (1988). *Science*, **241**, 53.
10. Maniatis, T., Fritsch, E. F., and Sambrook, J. (ed.) (1982). *Molecular Cloning*. Cold Spring Harbor Laboratory, Cold Spring Harbor, New York.
11. Hanahan D. (1985). In *DNA Cloning—A Practical Approach* (ed. D. M. Glover), Vol. 1, pp. 109–36. IRL Press, Oxford.
12. Hanahan, D. (1983). *Journal of Molecular Biology*, **166**, 557.
13. Carter, P., Bedouelle, H., and Winter, G. P. (1985). *Nucleic Acids Research*, **13**, 4431.
14. Caruthers, M. H. (1985). *Science*, **230**, 281.
15. Miller, J. H. (1972). *Experiments in Molecular Genetics*. Cold Spring Harbor Laboratory, Cold Spring Harbor, New York.
16. Birnboim, H. C. and Doly, J. (1979) *Nucleic Acids Research*, **7**, 1513.
17. Bankier, A. T. and Barrell, B. G. (1989). In *Nucleic Acids Sequencing—A Practical Approach* (ed. C. J. Howe and E. S. Ward), pp. 37–78. IRL Press at Oxford University Press, Oxford.

10

Cassette mutagenesis

JOHN H. RICHARDS

1. Introduction

Cassette mutagenesis provides a powerful and versatile approach for the generation of mutations and allows a wide range of studies of the relationship between base sequence and function in nucleic acids or between structure and function in proteins. The central aspect of this approach is the replacement of a region of DNA with new sequences that can be of any length, any sequence or any mixture of sequences. Thus, in studying the basis for protein function one can introduce not only changes of a few amino acids (an important aspect of primer-extension *in vitro* mutagenesis) but can also create chimeric proteins in which entire domains of a protein have been replaced with a region of totally different amino acid sequence. Examples of the latter include α-helical switches (1) between two DNA binding proteins, the cro protein and 434 repressor of bacteriophage 434. Substitution of the putative recognition helix of the 434 repressor with the putative recognition α-helix of 434 cro protein created a chimeric repressor whose specific DNA contacts are like those of 434 cro protein. Changes in catalytic specificity have also been introduced (2) into a β-lactamase by creating a chimera containing a region of an evolutionarily related penicillin binding protein of *Escherichia coli*, PBP-5, that has activity as a carboxypeptidase; this activity is completely lacking in the wild-type β-lactamase. The resulting chimera showed 1% of the carboxypeptidase activity of PBP-5.

One of the greatest powers of cassette mutagenesis stems from its ability to create families of genes that encode many different proteins that one can subsequently select or screen on the basis of function. Thus cassette mutagenesis allows the creation of specific mutants or the creation of families of mutants in which the differences in sequence are concentrated in targeted regions of the gene of interest.

To replace the desired region of a gene with the new cassette requires that the original gene contain restriction sites that allow cleavage of the gene at sites that flank the location for insertion of the new DNA. For greatest control and convenience these restriction sites should be unique to the construct being cleaved; the two sites should not be compatible with each other, though one (but not both) may be blunt-ended. The insertion of the cassette need **not** lead to the reconstitution of the sites used to open the vector in the step preceding insertion.

Cassette mutagenesis thus provides an *in vitro* strategy for the preparation of mutants that nicely complements primer-initiated site-directed mutagenesis; indeed, this latter technique is often used to introduce conveniently placed specific restriction sites (if not advantageously present) that are subsequently cleaved to allow the vector to accept the new DNA cassette.

A description of the experimental aspects of this approach falls naturally into a few discrete steps that will be described sequentially. The possible variations at each of these steps will be emphasized with particular stress on the significance of these variations for the types of approaches to structure/function studies they allow.

2. Requirement for restriction sites

Restriction sites in the gene should be localized as close as convenient to the segment that will be replaced. If this is a coding region, one examines the nearby sequence for places where base changes can be introduced that will lead to restriction sites generally **without** altering the amino acid sequence encoded. Such sites can usually be identified so that the cassette to be inserted does not exceed 100 base-pairs in length. A blunt-end site on one flank can be tolerated but blunt ends on both flanks should be avoided as they allow insertion of the new cassette in either orientation as well as greatly reducing the efficiency of incorporating the cassette.

Cases may arise in which other objectives can be realized when the insertion of the new cassette destroys the restriction site used to open the original vector. For example, after insertion of such a cassette any undesirable wild-type vector that formed by religation of the original fragments could be linearized by restriction digestion and subsequently easily separated from the desired circular product. Indeed the two processes, cutting and ligation could be carried out concurrently in a mixture of vector and cassette together with restriction endonucleases and ligases, because the desired product, after incorporation of the new cassette, would be resistant to further digestion.

As noted previously, unless appropriate restriction sites happen to be uniquely present in the native gene, they can most conveniently be introduced by primer-initiated site-directed mutagenesis, a subject which is covered in Chapters 1 to 5. Often to achieve the desired uniqueness of a given site for cassette mutagenesis, another site, with a similar recognition sequence, needs to be removed elsewhere in the gene or vector. If this site occurs in a structural protein, care must again be taken not to change the coding sequence; if the site lies in a control element one should be cautious that the change does not alter significantly an important function.

3. Preparation of vector to receive a cassette

After identification and/or creation of appropriate restriction sites, the vector is cleaved at these two sites for removal of the DNA segment to be replaced. Purification of the resulting vector (cleaved at two sites) from DNA that has been cut only once can be difficult as the segment to be removed is not generally longer than 100 base-pairs while the entire vector will probably be a few kilobases in length. This small fractional difference in size between the doubly and singly cut DNA complicates purification of the desired vector. If this singly cut DNA has not been removed it can regenerate the original double-stranded, circular DNA by religation leading to a background of wild-type genes as contaminants to the desired mutants.

A useful strategy to overcome this possible difficulty is shown in *Figure 1* and involves creation of the new construct by ligation of three fragments of

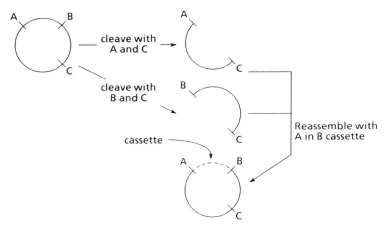

Figure 1. Strategy for the reconstruction of a plasmid containing a cassette. The approach depends on the three-way ligation of doubly digested vector fragments and the cassette. The vector fragments can easily be separated from uncut vector which might otherwise lead to a high wild-type background.

DNA (3). The vector is cleaved into two discrete fragments each obtained by double cutting (the two fragments should not be the same length). The desired fragments can be easily separated from any incompletely cut product so that the eventual three fragment reassembly of two segments of vector plus cassette will regenerate none of the original DNA.

Another approach for generating pure doubly cut vector uncontaminated by any singly digested plasmid, is to use a precursor vector with a spacer fragment of several kilobase-pairs inserted between the two restriction sites (4). In this situation the difference in size between the singly and doubly cut vectors ensures their ready separation.

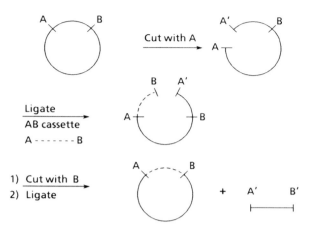

Figure 2. Two-step restriction digestion and ligation approach to clone a cassette.

There is a further possible strategy shown in *Figure 2* in which the vector is cut only once in the first stage, and the insert is ligated. Digestion with the second endonuclease is then followed by ligation of the previously inserted sequence to yield a circular product (5).

Finally, the use of endonucleases whose recognition sequence and sites of cutting are separated by intervening bases can allow cutting with an enzyme to generate a vector with two *non*-complementary protruding ends to receive the cassette. Examples of this tactic include the use of *Fok*I (6) and of *Bsp*MI (7) as illustrated in *Figure 3*.

The background of wild-type genes that might leak through the cutting and assembly steps can also be virtually eliminated if a restriction site, present in the wild-type gene has been removed in the newly created vector. In such situations, digestion of the ligated, duplex circular plasmids with the restriction endonuclease, whose recognition site has been removed during the mutagenesis (8), will linearize any remaining wild-type gene thereby greatly reducing its ability to transform the intended host bacterium (see Section 6 and *Protocol 5*).

3.1 Restriction digest of plasmid DNA

When digesting plasmids, a two- to tenfold excess of restriction endonuclease (units/μg DNA), depending on the enzyme efficiency, is generally needed for complete digestion. Determine the proper buffer for the enzyme as recommended by the manufacturer. When two different enzymes requiring different salt concentrations are to be used, it is best to do the reaction at low salt concentration, followed by that at high salt.

The volume of enzyme solution used should never go above **10%** of the

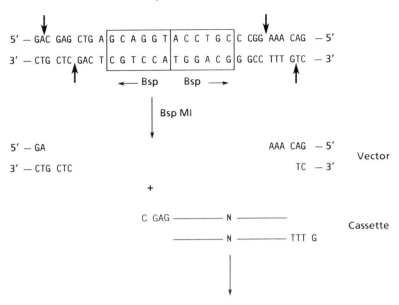

Figure 3. Generation of non-complementary ends by the use of the unusual restriction enzyme *Bsp*MI. This strategy allows vector molecules to be digested but produces non-complementary ends so the vector is unable to recircularize. A cassette with suitably designed ends can be ligated to the vector.

total mixture volume to insure against 'star' activity of the endonuclease. Conditions for a typical restriction enzyme digestion are given in *Protocol 1*.

Protocol 1. Restriction enzyme digestion

1. Typical restriction digest conditions:
   2 μl pBR322 (1 μg/μl)
   1 μl *Eco* RI (10 units/μl)
   2 μl H buffer (50 m Tris-HCl (pH 7.5), 10 mM MgCl$_2$, 100 mM NaCl, 1 mM DTT)
   14 μl H$_2$O

2. Incubate reaction at 37°C for 1 h.

3. Analyse DNA on a 1.2% agarose gel.

---

4. Preparation of cassette for insertion

The power and versatility of cassette mutagenesis derives largely from the essentially limitless variety of cassettes that can be prepared. One can create inserts either with specific base sequences or with mixtures of bases at any desired locations along the insert giving rise to families of mutants that allow very broad searches for those protein structures or those nucleic acid sequences that have the ability to perform some mandated function. In the most advantageous cases this function will affect the phenotype of the host organism in a way that permits efficient screening or selection.

4.1 Discrete sequence

The introduction of a cassette with a discrete base sequence allows creation and study of the particular protein or gene sequence thereby encoded. The basis for the construction of a chimeric protein will, for example, derive from design criteria based on the folding and functional properties of the constituent parts of the chimera. Though of central importance to such a construct these design considerations lie outside the scope of this chapter.

Experimentally the most straightforward procedure for the preparation of such specific cassettes is to chemically synthesize each of the two strands by the solid phase methods that can now be carried out automatically on a variety of commercially available machines designed for these purposes. The various manufacturers of the machines provide detailed instructions for their operation; the appropriate protected, activated bases are also available commercially.

The most common method of synthesis involves use of phosphoramidite chemistry (9). The single stranded oligonucleotides are purified by polyacrylamide gel electrophoresis, kinased, and annealed as described in *Protocol 2*. The annealing of two synthesized strands allows the resulting duplex cassette to have any desired ends: 5'-overhang, 3'-overhang, or blunt-end.

Protocol 2. Oligonucleotide synthesis purification, kinasing and annealing

A. Synthesis and purification

1. Synthesize the oligonucleotides (0.2 μmol) using phosphoramidite chemistry on an Applied Biosystems model 380A DNA synthesizer or other commercially available instrument. Dry down under vacuum.

2. Resuspend the oligonucleotides (⅓ of synthesis) in 25 μl of TE (10 mM Tris (pH 8.0), 1 mM EDTA) and 25 μl Maxam–Gilbert loading buffer [80% (w/v) deionized aqueous formamide, 50 mM Tris-borate (pH 8.3), 1 mM EDTA, 0.1% (w/v) xylene cyanol, 0.1% (w/v) Bromophenol Blue (10)] and purify the mixture on a preparative 15% polyacrylamide gel (11) (0.1 × 20 × 40 cm; 500 V for 12 h).

3. Remove the gel from the glass plates and wrap in celophane (e.g. Saran Wrap). Visualize the DNA by placing the gel on a fluorescent TLC plate and irradiating with short-wave UV. The DNA band appears as a shadow on the plate.

4. Excise the desired band, crush the gel, and suspend in 1 ml 2 M NaCl for 12 h at 37°C (or 4 h at 60°C).

5. Pass the solution through two Sephadex G-25 spin columns.

6. Alternatively, purify the oligonucleotides prior to trityl group removal by using the Applied Biosystems oligonucleotide purification cartridge (OPC) system (12).

B. Kinasing synthetic oligonucleotides

7. Add the purified oligonucleotides (100 pmol each) to $3 \mu l$ 1 M Tris (pH 8.0), $1 \mu l$ 100 mM DTT, $3 \mu l$ 10 mM ATP/100 mM MgCl$_2$, and 2–5 units of T4 polynucleotide kinase in a total volume of $30 \mu l$.

8. Incubate the reaction for 30 min at 37°C, followed by 10 min at 65°C to inactivate the enzyme.

C. Annealing synthetic oligonucleotides

9. Mix complimentary oligonucleotides (100 pmol each) with $10 \mu l$ 10× medium salt buffer (10 mM Tris–HCl (pH 7.5), 10 mM MgCl$_2$, 50 mM NaCl, 1 mM DTT) in a total volume of $100 \mu l$.

10. Heat to 90°C and allow the temperature to cool to room temperature slowly. It is best to immerse the mixture in its Eppendorf tube in a larger vessel (2–4 litres) for the annealing so that the temperature drop is gradual.

11. No further purification is necessary before ligation.

The cassette can, of course, be constructed from several oligonucleotides with complementary overlapping ends along the same lines that one uses in gene synthesis. This of course allows the assembly and insertion of cassettes much longer than would be feasible if one were restricted only to oligonucleotides that could be prepared by a single stepwise synthesis.

An element of variability in the resulting cassettes can be introduced by the shot gun ligation of overlapping synthetic oligonucleotides yielding double-stranded cassettes more than 120 nucleotides in length, as has been done in studying the enhancer region of the SV40 genome (13).

4.2 Multiple sequence cassette

A variety of synthetic strategies allow cassette mutagenesis to provide families of mutant genes that will endow their host organisms with different phenotypes that can be selected or screened on the basis of a function conferred by a particular wild-type or mutant gene. These approaches allow the rapid assessment of the properties of thousands or millions of different base or amino acid sequences. Very importantly, they allow the diversity being assessed to be unambiguously directed toward a particular region (or regions) of the gene or protein thereby avoiding the presence in the population of a high background of unsought mutations located outside a region of interest.

Moreover, the synthetic approach allows the creation of families of mutants with a variety of changes localized to the regions of interest. At one extreme is the method that inserts, into the family of mutants, every possible amino acid at one or more sites (3), 'site saturation'. In this strategy there remains no bias toward the wild-type sequence in the region thus mutagenized. However, the number of possible mutations increases rapidly, by $(20)^n$ where n is the number of amino acids saturated or by $(4)^n$ where n is the number of bases in a stretch of DNA saturated with all four possible bases. If one assumes a Poisson distribution of mutants and desires to ensure with a 95% confidence limit that all combinatorial possibilities have been tested in the eventual determination of function, one must assess about three times as many original transformants as there are possibilities.

In this, as in the other strategies for generating families of mutants, one should carefully plan the conceptual experimental approach. As has been documented in a study of the consensus sequence of an *E. coli* promoter, one can, by over mutagenizing, generate results that are too complex and vague to be practically useful whereas dissection of the problem into individual elements will greatly facilitate analysis of the mutational data (14).

4.2.1 Spiked oligonucleotides

In an alternative approach to site saturation one generates random mutations using low levels of nucleotide derivatives in each step of the chemical synthesis of the cassette. The result is to introduce point mutations spread throughout the region of interest. In this strategy the mutations will be present in an otherwise wild-type background, so that the sequences thus generated will not differ grossly from that of the parent. Accordingly, this approach on the one hand explores less of the possible amino acid space than does saturation with all twenty amino acids but allows mutagenic study of a larger region of a protein or genome without generating an indigestible number of mutants.

The level of mutation to be expected in any given situation can be calculated from the following relationship (15–17):

$$P(x) = N!)[x!(N-x)!]^{-1}C_m x[1 - C_m]N - x$$

where $P(x)$ equals the probability of x mutations per oligonucleotide of length N in a synthesis in which C_m is the concentration of contaminating nucleotide(s) relative to the concentration of wild-type nucleotide at each position.

As described in Chapter 9 an oligonucleotide spiked with mutagenizing bases as just described has been used (17) in a primer directed *in vitro* mutagenesis to study possible improvements in the catalytic potency of a mutant of triosephosphate isomerase previously significantly disabled by a

Glu 165 Asp mutation. In this work, screening of 150 000 transformants with mutagenic 75-base primers at a total contamination level of 2.5% of the three wrong phosphoramidites gave greater than 99% of the one-base changes and about 75% of the two-base changes that were possible in that oligomer window.

4.2.2 Mixed oligonucleotides

Between these two extremes (saturation at sites with twenty amino acids or low levels of mutant nucleotides randomly distributed in a longer stretch of gene) lie strategies in which all four bases might be incorporated at *one* position in a sequence of codons, giving thereby access to a subset of amino acid variants at those positions. In one of the earliest cases of cassette mutagenesis (8), subsets of amino acids were introduced into position 222 of subtilisin. A collection of these subsets contained all 20 amino acids at this site.

In an example of more extensive exploration of amino acid space, the sequence and structural requirements of a mitochondrial import signal were studied (18). The following mutagenic cassette was inserted with four bases equally incorporated at the first or second position in individual codons. The sites of substitutions are underlined, and the possible amino acids thereby encoded listed below each codon.

Wild-type	5′ — ATG-<u>G</u>TT <u>C</u>TA <u>C</u>CA A<u>G</u>A <u>C</u>TA <u>T</u>AT A<u>C</u>A GC<u>T</u> A<u>C</u>A A<u>G</u>T C<u>G</u>T GC<u>T</u> G<u>C</u>T— 3′

	M	V	L	P	R	L	Y	T	A	T	S	R	A	A
Potential amino acid substitutions		A	P	L	I	P	D	I	D	I	C	M	D	D
		D	Q	Q	T	Q	H	K	G	K	G	L	G	G
		G	R	R	K	R	N	R	V	R	R	P	V	V

Lastly, inosine has been inserted randomly into a cassette (in both strands) to direct mutagenesis with the findings amongst others that

(a) the frequency of inosine induced mutations was significantly less than that predicted from its content in the oligonucleotides, and

(b) inosine incorporation resulted almost exclusively in base changes to guanine (19).

In the synthesis of mixed oligonucleotides the incorporation of adenosine, thymine, guanine, cytosine, or inosine phosphoramidites (whether methyl or cyanoethyl) differs negligibly from their mole fraction in the mixture used in a coupling step. However, the stability of the deoxyguanosine methyl phosphoramidite in the solution is significantly lower than of the other derivatives so that, to achieve the expected incorporations, solutions of the mixed reagents should be prepared from solid phosphoramidites as soon as possible before their use in synthesis. This relatively more rapid rate of decomposition of the deoxyguanosine phosphoramidite is attenuated with cyanoethyl reagents due to their overall slower rate of decomposition (20).

For synthesis of mixed oligonucleotides, the solution containing the appropriate mixture of phosphoramidites should be prepared soon before use by dissolving the appropriate reagents in acetonitrite to give final concentration 0.1 M (this is the same concentration as used for incorporation of single bases during synthesis). If one desires to obtain not only substitutions at various positions, but also inserts of varying length, the capping step at each of the central stages of the synthesis can be omitted (22).

The various synthetic approaches outlined above enjoy a distinct advantage over mutations introduced by chemical or physical mutagens as the nature of the changes can be designed to explore any particular question of interest. For example, if one wishes to assess the functional properties of a particular helix in a complex protein, only those residues projecting from one face of the helix may be interesting sites for substitutions. For the reasons just described, chemical synthesis of an oligonucleotide cassette allows this objective to be realized so that one does not have to search for potentially interesting mutations (those confined to one side of one of the helices) against an overlarge background of probably useless changes (those in residues not involved in function but important in helix packing and therefore more likely to influence primarily folding or stability as distinct from function).

Lastly, the synthetic approach to random mutagenesis allows one to explore amino acid changes that would not be available from the use of random chemical mutagenesis. For example, the chemical mutagen N-methyl-N'-nitro-N-nitrosoguanidine (NTG) causes transitions (AT base-pairs changed to GC and vice versa) rather than transversions (AT base-pairs changed to TA or CG base-pairs) (21). Together with the low probability that more than one base in a particular codon is likely to be altered, this severely restricts the new amino acids to which such mutagenesis might give rise. Thus, this mutagen operating on a codon for alanine (GCX) would lead only to Thr(ACX) or Val(GTX).

4.3 Preparation of double-stranded cassettes

The preparation of the second strand of the cassette can be accomplished either by chemical synthesis or by biochemical approaches.

For a cassette that encodes a single DNA sequence, chemical synthesis clearly provides the method of choice. The two strands, both obtained by synthesis, can be annealed to generate a duplex cassette that can have ends of any desired type (5' or 3' overhangs or a single blunt end) for subsequent ligation into the vector (see *Protocol 2*, part C).

For cassettes composed of mixtures of sequences the tactics are more complex. A straightforward approach is again to prepare the second strand by total chemical synthesis with mixed bases incorporated at positions in the second strand complementary to those in the first. Careful annealing can then be used to form duplexes that will undoubtedly contain some fraction of mismatches at the positions of the mixed bases with possible complications arising from the subsequent *in vivo* editing that may occur and that could lead to families of mutants that do not contain the bases in the ratios introduced in the original chemical synthesis. Such a strategy also becomes more problematic with an increase in the number of sites being mutagenized. In spite of these potential problems, this approach has proved successful in site saturation studies of single amino acid residues in which the eventual appearance of various bases at the mixed positions closely approximated the ratios expected from synthesis (3, 22).

The preparation of a double-stranded cassette from chemically synthesized oligonucleotides randomly mutagenized at many positions throughout 270 bp of coding sequence has also been accomplished by directly annealing twelve overlapping oligonucleotides ranging from 30 bases to 67 bases in size. In this case the oligonucleotides enjoyed approximately 30 bp overlaps in relation to the opposite strand in order to minimize any bias against mutations in the overlapping regions (24).

A useful variant in the approach of direct annealing involves the incorporation, during chemical synthesis, of inosine (which can form base-pairs with all four normal bases) into the second strand at those sites complementary to the positions that contain mixed bases in the first strand (25).

A second general approach, illustrated in *Figure 4* involves the synthesis of the second strand by *in vitro* enzymatic techniques, thereby ensuring the incorporation into the second strand of bases complementary to those in the first strand and generating a duplex without base mismatches. The direct use of such *in vitro* synthesis restricts one to a duplex with a 3'-overhang and a blunt end and is accomplished by annealing a synthetic primer to the 3'-end of the first strand (leaving a 3'-overhang) followed by enzymatic extension to complete the second strand (which will necessarily have a blunt end at the 5'-end of the chemically synthesized first strand).

To avoid this limitation requires that the primer extension be carried

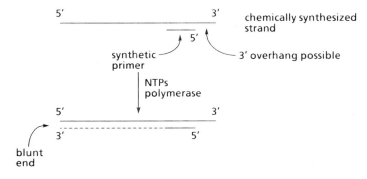

Figure 4. Generation of the second strand of a cassette by primer-directed DNA synthesis *in vitro*.

beyond the location of the desired restriction site so that subsequent digestion with a restriction endonuclease will create an overhang with any desired characteristics for subsequent ligation. In fact in this strategy the first strand is commonly synthesized with a few additional bases at both the 5'- and 3'-ends to increase the efficiency of endonuclease digestion.

An example of this approach is shown in *Figure 5* and involves the creation of a *Hind*III site and a *Sal*I site at either end of the cassette (18).

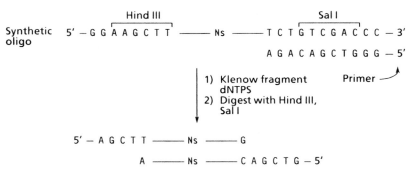

Figure 5. Creation of a cassette with overlapping ends following second-strand DNA synthesis *in vitro*.

A particularly elegant technique for preparation of cassettes with mixed bases is provided in *Protocol 3* and involves the chemical synthesis of a single strand that has a mutually complementary sequence at its 3'-end. This allows this strand (with as many mixed bases in its interior as desired) to hybridize with itself so that subsequent *in vitro* polymerase action will extend two new strands from the site of this hybridization, creating thereby a construct that is essentially double the length of the final cassette. Subsequent digestion with the two appropriate endonucleases generates therefore two moles of the final cassette from each mole of the intermediate construct. Any desired kind of terminus can be created by this versatile method (14, 23, 26).

An example of this approach illustrated in *Figure 6* involved the mutually primed synthesis from a *Sac*I site to which one additional nucleotide had been added to create an 8 bp palindrome. At the other end, which was destined to become a *Bam*HI site, three nucleotides were added 5' of the site to ensure greater cleavage efficiency and to discourage self-hybridization of this end of the synthetic oligomer (26).

Protocol 3. Cassette generation by mutually primed synthesis

1. Mix degenerate oligonucleotides (1–5 μg) in 10 μl 3× reaction buffer[a] and anneal by incubating at 37°C for at least 1 h before allowing the temperature to cool to room temperature slowly.

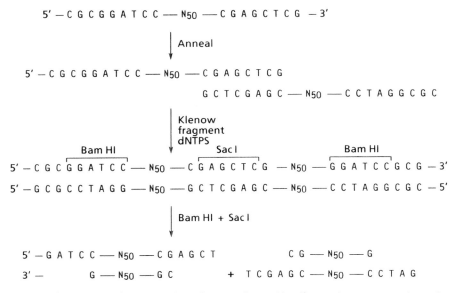

Figure 6. Generation of cassette dimer intermediates. 3' self-complementary ends on the single-strand cassette template are able to anneal. Synthesis of the second strand *in vitro* by mutual self-priming produces a cassette dimer that can be resolved to two copies of the cassette by suitable restriction digestion.

2. Add deoxynucleotide triphosphates (250 μM each of the four bases) and 10 μCi α-[^{32}P] dATP to the annealed oligonucleotides and adjust to a final volume of 30 μl.

3. Add Klenow fragment of *E. coli* DNA polymerase I (5 units) and incubate at 37 °C for 1–2 h.

4. Extract the reaction mixture with 60 μl of TE equilibrated phenol and ethanol precipitate the DNA.

5. Resuspend the DNA in 20 μl of TE and digest under appropriate buffer conditions with 10 to 50 units of restriction enzyme corresponding to the 5'-end site.

6. Phenol extract and ethanol precipitate, then isolate the synthetic cassette from a native polyacrylamide gel (see *Protocol 2*). In this case identify the product by autoradiography.

7. Cleave the purified DNA with the restriction endonuclease corresponding to the internal (original 3'-end) site.

8. Phenol extraction and ethanol precipitation. The synthetic cassette is now ready for ligation.

^{a} 3× reaction buffer is 30 mM Tris–HCl (pH 7.5), 30 mM MgCl$_2$, 150 mM NaCl, 15 mM DTT, 0.1 mg/ml gelatin).

In a fundamentally different approach (R. M. Fox, pers. commun.) the second strand of the cassette is **not** completed during the *in vitro* steps. Rather 'bracers' five to ten nucleotides in length are chemically synthesized and annealed at either end of the full length single-strand to form sites appropriate for subsequent ligation (these sites can be of any nature, 3'- or 5'-overhang or blunt-ended).

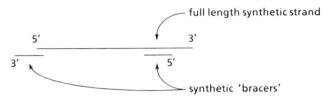

Figure 7. Illustration of the use of short 'bracer' oligonucleotides to form terminii suitable for ligation into a vector of the otherwise single-stranded cassette. Second-strand synthesis is carried out by the host cell.

After ligation and transformation of *E. coli*, the normal repair functions of the bacterium complete the second strand. In this procedure the transformation efficiency is somewhat lower, and the error rate somewhat higher than with completely double-stranded vectors but the ease of the procedure makes it potentially very attractive for general use.

5. Ligation

To assemble the final construct, one ligates the cassette, prepared by any of the various approaches outlined in Section 4.3, and generally with its 5'-ends phosphorylated, into the desired vector, which may concurrently be assembled from two fragments as discussed in Section 3 to ensure the absence of any parent DNA in the resulting preparation.

One commonly uses about a 10-fold excess of the cassette relative to the vector. If both restriction sites have cohesive ends, the total DNA concentration should be about 20×10^{-9} M (2×10^{-9} M vector, 20×10^{-9} M insert) whereas when one of the sites is blunt-ended the total DNA concentration should be about two times higher (one still uses about a 10-fold excess of cassette to vector). Even with these concentration differences, ligation involving one blunt and one cohesive end is considerably less efficient (by a factor of roughly 10) than ligation involving cohesive ends. The most commonly used enzyme is T4 DNA ligase which requires ATP as a co-factor (typically 1 mM). The amount of enzyme used varies almost as a matter of personal taste from somewhat less than 1 unit to 5 units in a 25 μl reaction volume though use of too high an amount of ligase may result in concatomers that do not subsequently transform the intended host. The ligation reaction is carried out as described in *Protocol 4A*. The unpurified vector/cassette ligation mixture is normally used directly to transform competent host cells such as *E. coli* (*Protocol 4B*).

If one uses a cassette whose 5′-ends have not previously been phosphorylated by treatment with ATP and an appropriate kinase, the amount of cassette relative to vector may be increased sometimes dramatically; in all events, the cassette should be in at least a ten fold excess. The resulting duplex will have two nicks, one in each strand at the sites of the 5′-ends of the cassette. The advantage of ligating unphosphorylated cassettes into the vector lies in the cassette's inability to join with each other to form multimers. A disadvantage is a somewhat lower transformation efficiency (about 50% that for completely covalent circular duplexes). For ligation of blunt ends, both vector and insert should have phosphorylated 5′-ends or the efficiency falls below useful levels.

6. Transformation

This step is carried out in the normal manner (see *Protocol 4*). For the preparation of families of mutants one should have careful regard for the number of cells in the original transformed sample that have acquired plasmid DNA as this represents the total number of plasmids that will have been sampled and will serve as the basis for knowing the probability that any given sequence will be tested in the subsequent screening. For example, if three times the number of possible variants are included in the original transformed sample, there is a probability of about 95% that any given sequence will be present in sample.

Protocol 4. Ligation and transformation

A. **Ligation**

1. Mix vector (0.04 pmol) and kinased/annealed cassette insert from step 8 of *Protocol 3* (0.4 pmol) with 5 μl 10 × ligase buffer^a and 1–5 units of T4 DNA ligase in a total volume of 50 μl. Note that the use of fresh DTT and ATP, and clean DNA (OD$_{260}$/OD$_{280}$ > 1.6) increases ligation efficiency (27).

2. Incubate the reaction at 15 °C for 15–18 h.

B. **Transformation**

3. Prepare competent *E. coli* by the Hanahan protocol (28).

4. Add an aliquot of the ligation mixture (1–10 μl) to 300 μl of competent cells. (It is best to do several transformations using different volumes of the ligation mixture because of the uncertainty of the ligation efficiency.)

Protocol 4. continued

5. Place the DNA/cell mixture on ice for 45 min, then heat shock at 42 °C for 90 sec.

6. Add 800 μl SOC medium[b] and incubate at 37 °C for 1 h to allow cell recovery.

7. Plate the cells onto the appropriate antibiotic-selection agar plates.

[a] 10× ligase buffer is 100 mM MgCl$_2$, 500 mM Tris (pH 8.0), 5 mM ATP, 50 mM DTT.
[b] SOC medium is 2% (w/v) bactotryptone, 0.5% yeast extract, 10 mM NaCl, 2.5 mM KCl, 10 mM MgCl$_2$, 10 mM MgSO$_4$ 20 mM glucose.

If insertion of a cassette results in the loss of a vector restriction site, as discussed in Section 3, the wild-type background can be significantly reduced by treating the ligation mix with the appropriate enzyme prior to transformation as described in *Protocol 5*.

Protocol 5. Cassette mutagenesis including selection by restriction digestion before transformation

1. Anneal complementary oligonucleotides (100 pmol each) exactly as described in step 9 of *Protocol 2*.

2. Ligate vector (0.04 pmol) and kinased/annealed cassette insert (0.4 pmol) as described in *Protocol 4*, part A.

3. Ethanol precipitate the DNA by adding 0.1 vol. 3 M sodium acetate, 2.5 vol. of absolute ethanol and place on dry ice for 15 min. Spin in a microcentrifuge (∼ 14 000 r.p.m.) for 15 min at r.t.

4. Wash the pellet with 100 μl of 70% ethanol, and dry briefly *in vacuo*. Resuspend in 20 μl TE.

5. To the DNA add 2 μl of the appropriate restriction enzyme reaction buffer and 1–10 units of the restriction endonuclease. This enzyme is the one that will cleave the site present in the parent plasmid but absent in the ligated cassette, thus linearizing any plasmid present as unwanted background and discouraging transformation by it.

6. Transform competent *E. coli* cells according to *Protocol 4*, part B.

References

1. Wharton, R. P., Brown, E. L., and Ptashne, M. (1984). *Cell*, **38**, 361.
2. Chang, Y.-H., Labgold, M. and Richards, J. H. (1990). *Proceedings of the National Academy of Sciences of the USA*, **87**, 2823.

3. Schultz, S. and Richards, J. H. (1986). *Proceedings of the National Academy of Sciences of the USA*, **83**, 1588.
4. Clarke, N. D., Lien, D. C., and Schimmel, P. (1988). *Science*, **240**, 521.
5. Chu, T. W., Grant, P. M., and Strauss, A. W. (1987). *Journal of Biological Chemistry*, **262**, 12806.
6. Vermesch, P. S. and Bennett, G. N. (1987). *Gene*, **54**, 229.
7. Stone, J. C., Vass, W. C., Willumsen, B. M., and Lowry, D. R. (1988). *Molecular and Cell Biology*, **8**, 3565.
8. Wells, J. A., Vasser, M., and Powers, D. B. (1985). *Gene*, **34**, 315.
9. Beaucage, S. L. and Carruthers, M. H. (1981). *Tetrahedron Letters*, **22**, 1859.
10. Maxam, A. M. and Gilbert, W. (1980). *Methods in Enzymology* (ed. L. Grossman and K. Moldave), Vol. 65, pp. 499–560.
11. Maniatis, T., Fritsch, E. F., and Sambrook, J. (ed.) (1982). *Molecular Cloning: A Laboratory Manual*. Cold Spring Harbor Laboratory, Cold Spring Harbor, New York.
12. Applied Biosystems Inc. (1989). *Catalogue*, Cat. No. 400771.
13. Grundström, T., Zeuke, W. M., Wintzerith, M. Matthes, H. W. D., Staub, A., and Chambon, P. (1985). *Nucleic Acids Research*, **13**, 3305.
14. Oliphant, A. R. and Struhl, K. (1987). *Methods in Enzymology* (ed. R. Wu). Vol. 155, pp. 568.
15. McNeil, J. B. and Smith, M. (1985). *Molecular and Cell Biology*, **5**, 3545.
16. Derbyshire, K. M., Salvo, J. J., and Gundley, N. D. F. (1986). *Gene*, **96**, 145.
17. Hermes, J. D., Blacklow, S. C., and Knowles, J. R. (1990). *Proceedings of the National Academy of Sciences of the USA*, **87**, 696.
18. Bedwell, D. M., Strobel, S. A., Yun, K., Jongeward, G. D., and Emr, D. S. (1989). *Molecular and Cell Biology*, **9**, 1014.
19. Nardinann, P. L., Markris, J. C., and Reznikoff, W. S. (1988). *Molecular and General Genetics*, **214**, 62.
20. Zon, G., Gallo, K. A., Samson, C. J., Shao, K. J., Summers, M. F., and Byrd, R. A. (1985). *Nucleic Acids Research*, **13**, 8181.
21. Birge, E. A. (1988). *Bacterial and Bacteriophage Genetics*, pp. 81–3. Springer-Verlag, New York.
22. Healey, W. J., Labgold, M. R., and Richards, J. H. (1990). *Proteins*, **6**, 275.
23. Oliphant, A. R. and Struhl, K. (1988). *Nucleic Acids Research*, **16**, 7673.
24. Murray, R., Pederson, K., Prosser, H., Muller, D., Hutchinson III, C. A., and Frelinger, J. A. (1988). *Nucleic Acids Research*, **16**, 9761.
25. Reidhaar-Olson, J. F. and Sauer, R. F. (1988). *Science*, **241**, 53.
26. Oliphant, A. R., Nussbaum, A. L., and Struhl, K. (1986). *Gene*, **44**, 177.
27. International Biotechnologies Inc. (1986/1987). *Catalogue*, pp. 87.
28. Hanahan, D. (1983). *Journal of Molecular Biology*, **166**, 557.

11

Recombination and mutagenesis of DNA sequences using PCR

ROBERT McARN HORTON and LARRY READINGTON PEASE

1. Introduction

The polymerase chain reaction makes possible a new method for engineering genes and is useful for both mutagenesis and recombination of sequences. This approach has significant advantages over current techniques because it is fast, straightforward, and not limited by the need for restriction sites.

1.1 The third generation of recombinant DNA technology

The development of recombinant DNA technology can be divided into three generations. The first generation approaches take advantage of natural recombination mechanisms *in vivo* to map and study genes. Into this category would fall classical genetic linkage analysis, and the use of mechanisms for homologous recombination between phages, for example, to generate sets of random recombinants between related genes. The second generation began with the use of restriction endonucleases and DNA ligase to specifically take genes apart, and put them back together in new combinations *in vitro*. It was the advent of this technology that made the field of molecular genetics what it is today.

Approaches based on restriction enzymes are limited, however, because they depend to some degree on good fortune to place appropriate restriction sites at convenient locations. The literature is replete with examples of gene constructs which include linkers, intron sequences, and long stretches of otherwise unnecessary DNA serving to connect restriction sites. In many cases these extra sequences are not a problem, but in other cases they may make a construct suboptimal, or even unworkable.

We are now witnessing the beginning of a third generation of recombinant technology, one in which specificity is provided by synthetic oligonucleotides rather than by restriction enzymes. The new generation does not depend on the occurrence of restriction sites and thus allows recombinations to be precisely made at essentially any position. This advance is made practical by the ready availability of custom-made synthetic oligonucleotides and by the polymerase chain reaction (PCR).

1.2 Synthetic applications of PCR

The polymerase chain reaction is widely used for **analysing** DNA sequences, but in fact it is essentially a method of **synthesizing** DNA *in vitro*. This synthetic process can be controlled to cause specific alteration or recombination of the newly made DNA.

In PCR, oligonucleotide primers are incorporated into the ends of the product DNA. The 5'-ends of these primers can contain any desired sequence, so long as the 3'-end matches the template well enough to prime (1). This means that a PCR product is chimeric, with the centre containing natural sequences copied from the template, and the ends containing synthetic sequences from the oligomers. Modification of the sequences in the oligomers leads to a very simple method of performing site-directed mutagenesis at the ends of a PCR product (see for example ref. 2). However, there are many instances where it would be desirable to introduce a mutation into the centre of a PCR product, at an arbitrary distance from the end. In that case, the mutation cannot be included in the PCR primer, because the primer would have to be very long to reach all the way into the centre of a reasonably large fragment. This limitation has been overcome by the use of a generally applicable PCR mediated recombination technique we call overlap extension.

2. Overlap extension

The concept of overlap extension allows PCR to be used for both introducing site-directed mutations into the centre of a fragment and for creating recombinant DNA molecules. When it is used for joining or 'splicing' different genes together, we call the process 'gene SOEing' or '"Splicing" by Overlap Extension'.

2.1 Mechanism

The process of overlap extension depends on the fact that sequences added to the 5'-end of a PCR primer become incorporated into the end of the product molecule. By adding the appropriate sequences, a PCR amplified segment can be made to 'overlap' or share sequences with another segment. In a subsequent reaction, the overlap serves as a primer for extension by DNA polymerase, which creates a recombinant molecule. An overview of the method and its applications is given in *Figures 1* to *3*. The mechanism is described in general terms below and specific examples are given in Section 3.

2.1.1 Two-sided overlaps

The first groups to develop overlap extension used 'two-sided' approaches in which the reaction is carried out symmetrically (3, 4). PCR is used to modify the ends of both of the fragments to be recombined so that they share

R. M. Horton and L. R. Pease

How Overlap Extension Works:

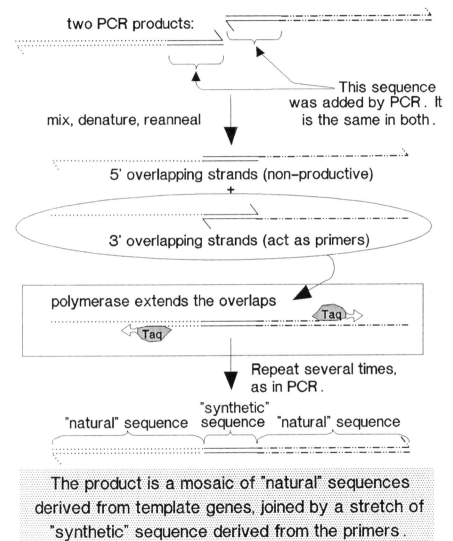

two PCR products:

This sequence was added by PCR. It is the same in both.

mix, denature, reanneal

5' overlapping strands (non–productive)

+

3' overlapping strands (act as primers)

polymerase extends the overlaps

Taq

Taq

Repeat several times, as in PCR.

"synthetic"
"natural" sequence   sequence   "natural" sequence

The product is a mosaic of "natural" sequences derived from template genes, joined by a stretch of "synthetic" sequence derived from the primers.

Figure 1. Outline of the principle of overlap extension PCR to SOE DNA to create a recombinant molecule.

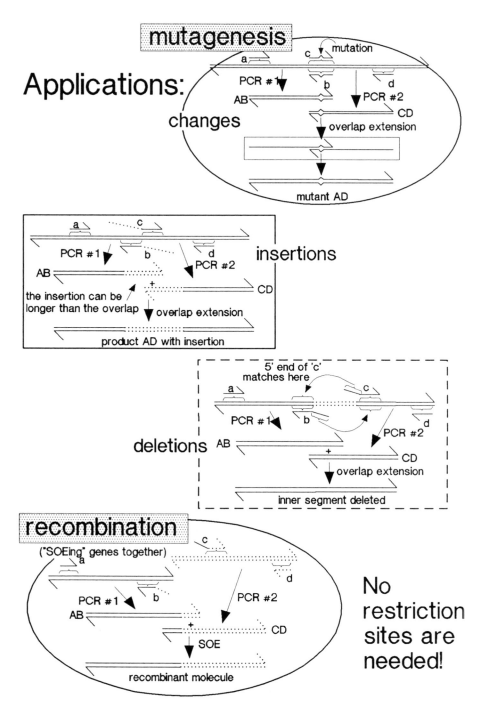

Figure 2. Outline of some current applications of overlap extension PCR for the mutagenesis and recombination of DNA molecules.

One Sided SOEing

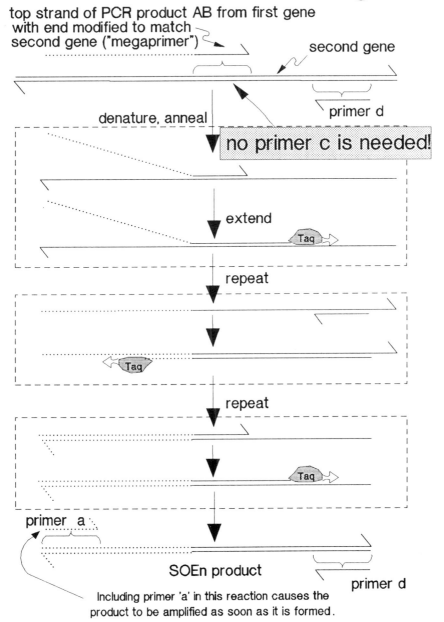

top strand of PCR product AB from first gene
with end modified to match
second gene ("megaprimer")

second gene

primer d

denature, anneal

no primer c is needed!

extend

Taq

repeat

Taq

repeat

Taq

primer a

SOEn product

primer d

Including primer 'a' in this reaction causes the
product to be amplified as soon as it is formed.

Figure 3. One-sided overlap extension.

sequences in common at the ends to be joined. This allows one strand of each fragment to 'overlap' with a strand from the other fragment when they are mixed, denatured, and reannealed (see *Figure 1*). One pair of strands overlap at their 5'-ends and are not productive. In the other pair, however, the overlap provides free 3'-ends which can serve as primers for DNA polymerase, which then extends the overlap in both directions to produce the recombinant molecule. This process occurs under the conditions of a PCR, in which there are multiple cycles of heat denaturation, reannealing, and extension. By including 'flanking' primers ('a' and 'd' in *Figure 2*), a concurrent PCR causes the recombinant product to be amplified in the same reaction.

2.1.2 One-sided overlaps

'One-sided' SOEing uses one less primer to achieve essentially the same results (*Figure 3*). This has also been called the 'megaprimer' method (5), because one strand of an end-modified PCR product is used as a very long primer in a subsequent reaction. This strand can be generated using asymmetric PCR (6), or it may be added as part of a double-stranded molecule, because the complementary (non-productive) strand does not seem to interfere with the reaction (see Section 2.4).

2.2 Mutagenesis

The concept of joining two DNA fragments together by overlap extension provides the key to introducing mutations into the centre of a PCR fragment, as shown in *Figure 2*. PCR is used to produce changes near the overlapping ends of the targeted fragments. When these fragments are combined in an overlap extension reaction the mutation is placed inside the product molecule, at a more or less arbitary distance from either end (3, 4). The one-sided overlap extension methods make this process more economical because only one new primer must be synthesized for each new mutation. A procedure for mutagenesis by overlap extension is given in *Protocol 1*. PCR solutions and reagents are listed in *Table 1*.

2.3 Recombination

This method of joining PCR amplified sequences can be applied to DNA sequences which originated from different genes (SOEing genes together). By incorporating sequences into the 5'-end of one primer to make it complementary to sequences in the other gene, it is possible to make any PCR product overlap with any other gene segment, so that they can be recombined. *Figures 1* and *2* illustrate how two gene segments can be SOEn together.

SOEing (see *Protocol 2*) can be performed in a one-sided manner, as with mutagenesis, but when two unrelated templates are used the process can be

---

Protocol 1 Mutagenesis by overlap extension

1. In separate tubes, assemble the reactants for PCRs #1 and 2 to make AB and CD.

PCR#	1	2
product	AB	CD
components: template	parental gene	parental gene
primer #1	'a'	'c'*
primer #2	'b'*	'd'
10 × buffer	10 μl	10 μl
10 × dNTPs	10 μl	10 μl
Taq polymerase	0.5 μl	0.5 μl
H$_2$O	to 100 μl	to 100 μl

* denotes mutagenic primers.

2. Cover the aqueous phase with several drops of mineral oil (Sigma). Cycle for 15 to 25 rounds (94°C for 30 sec, 50°C for 2 min and 72°C for 1 min).

3. Run the reactions on an agarose gel. Cut out the bands that are the right size for AB and CD.

4. Recover the DNA from the bands. Electroelution, GeneClean, or freeze-squeeze (*Figure 9*) may be used.

5. In a new tube, mix the reactants for the overlap extension reaction, and subject to thermal cycling as in step 2 above.

Overlap extension reaction

product	AD
Components: template #1	AB
template #2	CD
primer #1	'a'
primer #2	'b'
10 × buffer	10 μl
10 × dNTPs	10 μl
Taq polymerase	0.5 μl
H$_2$O	to 100 μl

Note: One-sided overlap extension ('megaprimer' mutagenesis) differs as follows:
(a) Only one initial PCR is done, to make product AB.
(b) Template #2 in the overlap extension reaction is now the parental gene itself.
(c) No primer #1 (primer 'a') is added to the overlap extension reaction (the top strand of AB takes its place – adding primer #1 would merely amplify the parental sequence).

Table 1. Solutions and reagents

***Taq* polymerase:** is available from Cetus corporation. It comes at a concentration of $5\,U/\mu l$. The enzyme can be purchased separately, or as part of a 'GeneAmp' kit, which includes buffers and positive controls.

10 × PCR buffer: (this is the buffer recommended by Cetus)

Stock solution	Amount	[1 X]	[10 X]
1 M KCl	$500\,\mu l$	50 mM	500 mM
1 m Tris–Cl, pH 8.3	$100\,\mu l$	10 mM	100 mM
150 mM MgCl$_2$	$100\,\mu l$	1.5 mM	15 mM
1% (w/v) gelatin	$100\,\mu l$	0.01%	0.1%
H$_2$O	$300\,\mu l$		
Total:	$1000\,\mu l$		

dNTPs: $1\times = 200\ \mu M$ $10\times = 2\,mM$ each dNTP.

Note: the dNTPs in a GeneAmp kit from Cetus are NOT at 10 × concentration; they come as 10 mM stocks ($50\,\mu l$ dATP + $50\,\mu l$d CTP + $50\,\mu l$ dGTP + $50\,\mu l$ TTP + $50\,\mu l$ H$_2$O = $250\,\mu l$ 10 × dNTPs). Convenient dNTP solutions are also available from Pharmacia. Salts of the dNTPs are also available in solid form, but these are much less convenient to use because they must be resuspended, brought to the proper pH, and the actual concentration must be determined by UV absorbance (see ref. 9, p. 449).

---

modified by including the flanking primers ('a' and 'd' in *Figure 3*). The recombinant product is then amplified by PCR as soon as it is formed, increasing the yield. This cannot be done in one-sided mutagenesis because primers 'a' and 'd' could amplify the parental sequences. (Note that in two-sided mutagenesis, the parental sequences are not present in the final reaction, so the flanking primers can be included).

Protocol 2. Gene SOEing

1. In separate tubes, amplify intermediate products AB and CD by assembling the above components.

PCR#	1	2
product	AB	CD
components: template	Gene I	gene II
primer #1	'a'	'c'*
primer #2	'b'*	'd'
10 × buffer	$10\,\mu l$	$10\,\mu l$
10 × dNTPs	$10\,\mu l$	$10\,\mu l$
Taq polymerase	$0.5\,\mu l$	$0.5\,\mu l$
H$_2$O	to $100\,\mu l$	to $100\,\mu l$

* denotes SOEing primers.

2. Cover with mineral oil, and cycle for 15 to 25 rounds (94°C for 30 sec, 50°C for 2 min and 72°C for 1 min).

3. Run the reactions on an agarose gel. Cut out the AB and CD bands.

4. Recover the DNA from the band. Electroelution, GeneClean, or freeze-squeeze (*Figure 9*) may be used.

5. In a new tube, mix the reactants for the SOEing reaction, and cycle as step 2 above.

SOEing reaction

	product	AD
components:	template #1	AB
	template #2	CD
	primer #1	'a'
	primer #2	'b'
	10 × buffer	10 μl
	10 × dNTPs	10 μl
	Taq polymerase	0.5 μl
	H₂O	to 100 μl

Note: One-sided SOEing is similar except:
(a) Product CD is not amplified from gene II; only AB is made in a separate PCR.
(b) Gene II takes the place of template #2 in the SOEing reaction.
(c) Unlike one-sided ('megaprimer') mutagenesis, both flanking primers ('a' and 'd') are added to the SOEing reaction. The only thing they should be able to amplify is the recombinant product.

2.4 What happens to the non-productive strands?

The mechanisms illustrated above focus only on the strands having the overlap at their 3'-ends, because these are the only strands which can be extended by polymerase, which always polymerizes from the 3'-end. People sometimes wonder what becomes of the other strand. We propose that there are two things happening to these 'non-productive' strands in an overlap extension reaction. First, if the flanking primers ('a' and 'd' in *Figure 2*) are included in the reaction, the non-productive strands can serve as templates to generate more of the productive strands. Second, if the non-productive strand hybridizes to the recombinant product, it will be downstream of the polymerase which is advancing from the flanking primer. This would make it susceptible to degradation by a 5' to 3' exonuclease activity of the polymerase. Though *Taq* polymerase apparently has no 3' → 5' exonuclease activity, it does have homology to the domain of *E. coli* DNA polymerase I which has 5' → 3' exonuclease activity (7). Thus, the non-productive strands may be degraded during the overlap extension reaction. At any rate, the reaction is not hindered by the presence of these strands (5).

3. Examples of overlap extension applications

3.1 Mutagenesis

Examples of the use of overlap extension for mutagenesis are illustrated in *Figure 4*. These examples show the two-sided approach in which two overlapping oligonucleotides were used for each mutation. Two separate PCRs were carried out on the template gene to give products AB and CD. These products each have the desired mutations incorporated into their ends. The final product AD is made by hybridizing the overlapping strands from the two fragments and extending this overlap with DNA polymerase. The mutant product AD contains no parental sequence because the reactant fragments AB and CD contained only the mutant sequence at the point of interest. The only copies of the wild-type sequence are in the original template molecules and these are removed before the overlap extension reaction. Thus this procedure should be 100% efficient in terms of generating mutant molecules with 0% wild-type leaking through. In one case, of 168 clones probed with oligonucleotides specific for the mutant sequence, only three did not hybridize, so 98% of the clones contained the desired mutations (4). The negative

```
(A) Single base substitution (from reference 4).
Parent gene (H-2Kᵇ):
   amino acid
   #:165 166 167 168 169 170 171 172 173 174 175 176 177 178 179
5'-...GTG GAG TGG CTC CGC AGA TAC CTG AAG AAC GGG AAC GCG ACG CTG...-3'
3'-...CAC CTC ACC GAG GCG TCT ATG GAC TTC TTG CCC TTG CGC TGC GAC...-5'
                                     *
        Primer 'b':      3'-t atg gac ctc ttg ccc-5'
                            |<-overlap region-->|
        Primer 'c':      5'-a tac ctg gag aac ggg-3'
                                     *
Product AB:                          *
5'-...GTG GAG TGG CTC CGC AGA TAC CTG GAG AAC GGG-3'
3'-...CAC CTC ACC GAG GCG TCt atg gac ctc ttg ccc-5'
                                     *
Product CD:                          *
                         5'-a tac ctg gag aac ggg AAC GCG ACG CTG...-3'
                         3'-T ATG GAC CTC TTG CCC TTG CGC TGC GAC...-5'
                                     *
```

Top strand of AB overlaps with bottom strand of CD:

```
5'-...GTG GAG TGG CTC CGC AGA TAC CTG GAG AAC GGG-3'
                         3'-T ATG GAC CTC TTG CCC TTG CGC TGC GAC...-5'
                                     *
```

Polymerase extends the overlaps to make the mutant molecule:

```
                                     *
5'-...GTG GAG TGG CTC CGC AGA TAC CTG GAG AAC GGG AAC GCG ACG CTG...-3'
3'-...CAC CTC ACC GAG GCG TCT ATG GAC CTC TTG CCC TTG CGC TGC GAC...-5'
                                     *
```

Figure 4. Examples of mutagenesis by overlap extension. This figure details the actual sequences of primers, templates, and various intermediates involved in mutagenesis by overlap extension. In part A, a two-sided approach is used to perform the simplest type of mutation, the single base-pair change. Here the mutagenic oligonucleotides 'b' and 'c' are entirely complementary, so that the overlap is as long as the primers. The altered nucleotide is marked by an asterisk. Only the region near the mutation site is shown. Ellipses indicate that a DNA strand extends beyond the segment illustrated. The flanking primers 'a' and 'd' are not shown. The synthetic DNA of the primers is represented by

clones were not further characterized but they were probably cloning arte-
facts rather than wild-type molecules. Conventional methods give efficiencies of 50 to 80% (8).

The overlap extension procedure, in common with most procedures for
site-directed mutagenesis, involves *in vitro* synthesis of parts of the mutant
molecule and is subject to random misincorporations of nucleotides. Thus,
the mutant product should be sequenced to be sure it is correct. We have
found an error frequency of 0.026% (1 mistake in approximately 4000
nucleotides sequenced) among clones we have made with the two-sided
mutagenesis strategy outlined above, which involves subjecting each segment
of the gene to two sets of PCR or overlap extension reactions (4). This
frequency is low enough to make the method practical. It is possible on
theoretical grounds that one-sided overlap extension approaches would
have even lower errors frequencies, because the molecule is subjected to
fewer sets of reactions (5). However, this remains to be demonstrated experimentally.

```
(B) Multiple base substitutions (from Hunt et al., in preparation).
Parent gene (H-2Kᵇ):
  amino acid
     #: 18   19   20   21   22   23   24   25   26   27   28   29   30   31   32
5'-...GGG  GAG  CCC  CGG  TAC  ATG  GAA  GTC  GGC  TAC  GTG  GAC  GAC  ACG  GAG...3'
3'-...CCC  CTC  GGG  GCC  ATG  TAC  CTT  CAG  CCG  ATG  CAC  CTG  CTG  TGC  CTC...-5'
                                *    * ***
Primer 'b':3'-ggg gcc aag tag aga cag ccg atg cac c-5'
                        ¦<---overlap region-->¦

Primer 'c':               5'-c tct gtc ggc tac gtg gac aac acg-3'
                              * ***                         *
Product AB:          *         *  * ***
5'-...GGG GAG CCC CGG TTC ATC TCT GTC GGC TAC GTG G-3'
3'-...CCC CTC ggg gcc aag tag aga cag ccg atg cac c-5'
                          *    * ***
Product CD:                    *  * ***                    *
                          5'-c tct gtc ggc tac gtg gac aac acg GAG...-3'
                          3'-G AGA CAG CCG ATG CAC CTG TTG TGC CTC...-5'
                              * ***                         *
```

Top strand of AB overlaps with bottom strand of CD:

```
                      *    * ***
5'-...GGG GAG CCC CGG TTC ATC TCT GTC GGC TAC GTG G-3'
                      3'-G AGA CAG CCG ATG CAC CTG TTG TGC CTC...-5'
                          * ***                         *
```

Polymerase extends the overlaps to make the mutant molecule:

```
                      *    * ***                    *
5'-...GGG GAG CCC CGG TTC ATC TCT GTC GGC TAC GTG GAC AAC ACG GAG...-3'
3'-...CCC CTC GGG GCC AAG TAG AGA CAG CCG ATG CAC CTG TTG TGC CTC...-5'
                      *    * ***                    *
```

lower-case letters, while DNA synthesized by polymerase is shown in capital letters.
Strands which have incorporated synthetic primers thus have lower case letters at the 5'-
end. Note that these are the 'non-productive' strands: the strands which go on to form the
mutant product are the ones with the 3'-ends shown. Part B shows the construction of a
more complicated mutant containing a total of six nucleotide substitutions. In this
example the overlap region shared by the two primers is shorter than either of the
primers themselves.

3.2 Recombination

As examples of the use of gene SOEing we show constructions of two chimeric genes in which the recombinations were made specifically at locations devoid of convenient restriction sites.

3.2.1 Promoter switching

For the first example, we detail the construction of a molecule in which the promoter and 5′ untranslated regions from a class II MHC molecule have been attached to a class I MHC gene, with the recombination placed precisely at the initiation ATG codon. This recombinant is the first SOEn molecule to be expressed in transgenic mice, and will be described in more detail elsewhere (Cai, Donovan, and Pease, unpublished data).

The general approach used in the construction is illustrated in *Figure 5*. The SOEn fragment includes a restriction site from the class I gene on its right end, and a site from the class II promoter region on its left. This allows it to be cloned into a position in a plasmid containing the remaining portions of the appropriate genes. The other portions of the construct were cloned into the plasmid in the conventional (second generation) manner using available restriction sites. Details of the site of gene SOEing are given in *Figure 6*.

3.2.2 A fusion protein

As another example of the use of gene SOEing, we show the construction of a gene for a fusion protein between an antigen (ova, chicken ovalbumin) and a selectable marker (neo, an enzyme conferring neomycin resistance). This gene was made as part of a project to establish a system for mapping the regions of protein antigens which are recognized by cytotoxic T-lymphocytes (L. R. Pease *et al.*, unpublished data). For construction of this fusion protein gene we took advantage of both one-sided and two-sided SOEing to make a three-part chimeric fragment using only four oligonucleotides, as shown in *Figures 7* and *8*.

Note that in both of the examples of gene SOEing given above only the recombination which needed to be at a precise position was accomplished by gene SOEing. Those which could be more flexibly located (i.e. anywhere in an intron, or in some non-vital region of a plasmid) were made using restriction sites.

4. Planning a strategy

4.1 Is PCR the solution to your problem?

In terms of efficiency and simplicity, mutagenesis by overlap extension is comparable or superior to most other approaches available. It is capable of generating insertions or deletions as well as changes (*Figure 2*). This approach is a good choice for beginners deciding on a technique to learn because all of

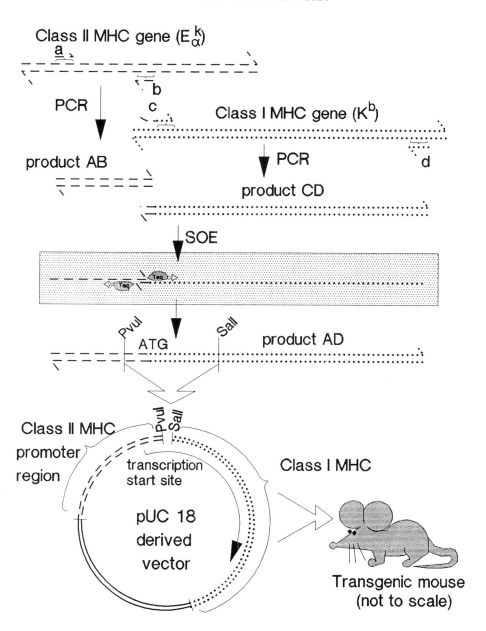

Figure 5. Recombination by gene SOEing, example #1: promoter switching. The recombination site was located precisely at the initiation methionine so that the entire protein-coding region is from the class I gene while the promoter and 5′ untranslated region are from class II. This construct has been successfully incorporated into a transgenic mouse. The sequences at the recombination site and of the primers used are given in *Figure 6*.

```
First gene (H-2Eₐᵏ):
5'-...GCGATCGCTTCTGAACCCACCAAACACCCAAGAAGAAA ATG GCC ACA...-3'
3'-...CGCTAGCGAAGACTTGGGTGGTTTGTGGGTTCTTCTTT TAC CGG TGT...-5'

Second gene (H-2Kᵇ):
5'-...CCGCAGAACTCAGAAGTCGCGAATCGCCGACAGGTGCG ATG GTA CCG TGC ACG CTG...-3'
3'-...GGCGTCTTGACTCTTCAGCGCTTAGCGGCTGTCCACGC TAC CAT GGC ACG TGC GAC...-5'


Primer 'b':                     3'-ttgtgggttcttcttt tac-5'
                                 |<-priming region->|
                                 |<-overlap region->|
                                                    |<-priming region-->|
Primer 'c':                     5'-aacacccaagaagaaa atg gta ccg tgc acg c-3'

Product AB (amplified from H-2Eₐᵏ):
5'-...GCGATCGCTTCTGAACCCACCAAACACCCAAGAAGAAA ATG-3'
3'-...CGCTAGCGAAGACTTGGGTGGTttgtgggttcttcttt tac-5'

Product CD:
                                5'-aacacccaagaagaaa atg gta ccg tgc acg cTG-3'
                                3'-TTGTGGGTTCTTCTTT TAC CAT GGC ACG TGC GAC-5'
```

The top strand of AB overlaps with bottom strand of CD:
```
5'-...GCGATCGCTTCTGAACCCACCAAACACCCAAGAAGAAA ATG-3'
                           3'-TTGTGGGTTCTTCTTT TAC CAT GGC ACG TGC GAC-5'
```

Polymerase extends the overlaps to make the recombinant molecule:
```
5'-...GCGATCGCTTCTGAACCCACCAAACACCCAAGAAGAAA ATG GTA CCG TGC ACG CTG-3'
3'-...CGCTAGCGAAGACTTGGGTGGTTTGTGGGTTCTTCTTT TAC CAT GGC ACG TGC GAC-5'
```

Figure 6. Details of the primers and sequences involved in the promoter switch example. This figure is similar to *Figure 4*. The coding regions of the genes are shown with spaces between the codons. With gene SOEing the primers can be described as having distinct 'priming' and 'overlap' regions. The priming region, at the 3'-end of the oligomer, allows it to act as a primer by binding to its template gene (in mutagenesis the entire oligomer acts as a priming region, except for the bases which are being changed). The overlap region contains sequences added to the 5'-end of a primer which makes it complementary to the other SOEing primer. In this case the entire overlap has been added to primer 'c' so that primer 'b' contains only sequences from the gene donating the promoter region (H-2Eₐᵏ). The two regions share three bases in common; these are from the codon encoding the initiation methionine which is the same in both genes. Note that the recombination is made exactly at the initiation codon even though there are no restriction sites at this location.

the technical expertise needed is useful for any PCR application and does not involve the use of any complicated prokaryotic genetics or esoteric bacterio-phage vectors which are useful for few other applications.

For purposes of recombination, splicing by overlap extension, being a third generation technology, has the significant advantage that it is not limited by the presence or absence of restriction sites. This makes it very useful for making recombinants between DNA segments which do not happen to have unique restriction sites in convenient locations. Moreover, mutagenesis and recombination can be carried out simultaneously. However, there are many cases in which the disadvantages outweigh the benefits. First, the size of a

DNA fragment which can be enzymatically manipulated *in vitro* by PCR is, for practical purposes limited, and it is usually convenient to have a cassette system with appropriate restriction sites for cloning the PCR modified fragments (see Section 4.3, below and Chapter 10). Second, for every recombination at least one new oligonucleotide (about 32 bases) must be synthesized. At current bargain basement oligo prices this would cost about $70 (US). Second generation (i.e. restriction enzyme based) approaches may be cheaper for routine use. Finally, since PCR-based manipulation of sequences involves the *in vitro* synthesis of the product by an enzyme that has been removed from its natural environment and asked to perform in a tube there is a possibility of error. Though in our experience the frequency of error is rather low (see Section 5.3) mistakes do happen and it is recommended that the cloned product be sequenced to find a clone without mistakes. This would be too much trouble for an ordinary everyday recombinant molecule.

Gene SOEing is a very attractive alternative, however, in those cases which are the most troublesome for standard second generation methodologies. A prime example is the generation of precise fusion proteins in which two segments of coding regions must be joined in frame without adding unwanted codons. In these situations it is extremely rare to find restriction sites exactly where they are needed, and often it may not even be possible to add restriction sites in the ideal positions without making an amino acid change. Whereas to use second generation technology in these situations might require a convoluted series of steps, the use of third generation technology remains straightforward.

4.2 Oligonucleotide design

A SOEing oligonucleotide can be described as having regions which serve different functions. The 3'-end of the oligo must contain sequences which allow it to act as a primer on its template; we call this the '**priming region**'. At its 5'-end it contains sequences which overlap with the sequence to be joined; we call this the '**overlap region**'. Between these regions a third '**insertion region**' can be included for insertional mutagenesis. Examples of SOEing oligos with their regions labelled are given in *Figures 6* and *8*.

4.2.1 Overlap design

The priming and overlap regions of the oligos used in these studies are both approximately 16 bp long. For some primers we have attempted to take into account the fraction of G's and C's in the sequence in determining the length of overlaps. For lack of a more appropriate formula, we have used the 'estimated T_d' (9) as a guide in deciding how long to make priming and overlap regions.

The estimated T_d is an empirically derived formula which usually provides a fairly accurate guess of the temperature at which an oligonucleotide will bind

Figure 7. Recombination by gene SOEing, example #2: a gene for a fusion protein. Part A diagrams the plan followed for this construction, which uses both one-sided and two-sided approaches. The primers are named by analogy to the two-sided method: there is no primer 'b' and no primer 'e'. The sequences used for the one-sided reactions (SOE reactions #1 and #2) are given in *Figure 8*. The overlap involved in the two-sided SOEing reaction (#3) is the size of fragment CD, 164 base-pairs. This serves to illustrate that such overlaps may be significantly longer than the 16 or so base-pairs usually used. Note that this construction is in effect an example of insertional mutagenesis (see *Figure 2*) in which the primers carrying the insertion are actually the strands of a PCR amplified product. Part B is a photograph of an ethidium bromide stained gel showing the various products and intermediates. Reaction conditions were as described in *Table 3*, using templates and primers as follows:

Reaction:	PCR	SOE #1	SOE #2	SOE #3
product	CD	AD	CF	AF
template #1	ova	neo	CD	AD
template #2	–	CD	neo	CF
primer #1:	'c'	'a'	'c'	'a'
primer #2:	'd'	'd'	'f'	'f'

Expected sizes of the bands are: CD = 164 bp, CF = 374 bp, AD = 451, AF = 661 bp.

---

to 50% of a target sequence on a solid support in a standard salt buffer (5 × SSPE). It is calculated as follows:

$$((\# \text{ G's} + \# \text{ C's}) \times 4) + ((\# \text{ A's} + \#\text{T's}) \times 2)$$

This number is the approximate denaturation temperature of the oligonucleotide in degrees Celsius. Our oligos are designed so that the priming and overlap regions have estimated T_d's of approximately 50°C. Thus, a primer having a high percentage of G's and C's can be shorter than one having mostly A's and T's.

One word of caution: the estimated T_d is not an accurate estimation of the annealing temperature of an oligo in PCR buffer which contains, among other things, Mg^{2+}, which is not included in 5 × SSPE. Also, one might be tempted to place the oligo in a very GC-rich region in order to achieve the 50°C temperature with as few residues as possible. To do so would violate a commonly accepted standard of primer design, which is to have a decent balance of nucleotides in the primers.

An alternative to calculating estimated T_d's is to simply make the overlap and priming regions about 16 bases long. The overlap can clearly be much longer than this as shown in *Figure 7* (SOE #3). We have not rigorously studied the minimum length (or estimated T_d) which can be used but 16 bp (or 50°C) works reproducibly.

Primer sequences should always be checked to be sure that they do not

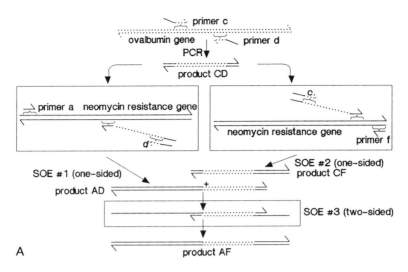

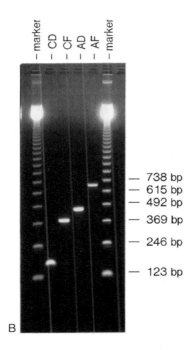

form any obvious hairpin loops, and that they do not hybridize to the other primer in the same reaction. It is also important that the most 3′ four or five bases match the template if a sequence is to act as a primer. Mismatches too close to the 3′-end will prevent priming.

(A) *Generating fragment AD by one-sided SOEing:*

```
ova:
amino acid
   #:      236 237 238 239 240 241 242 243 244 245 246 247 248 249
   5'-...AGT GGG ACA ATG AGC ATG TTG GTG CTG TTG CCT GAT GAA GTC...-3'
   3'-...TCA CCC TGT TAC TCG TAC AAC CAC GAC AAC GGA CTA CTT CAG...-5'

Primer 'c':
         5'-ctt ctt gac gag ttc ttc ttg gtg ctg ttg cct gat-3'
            |<----overlap region--->|<---priming region-->|

Product CD (left end):
         5'-ctt ctt gac gag ttc ttc ttg gtg ctg ttg cct gat GAA GTC...-3'
         3'-GAA GAA CTG CTC AAG AAG AAC CAC GAC AAC GGA CTA CTT CAG...-5'
```

The bottom strand of CD acts as a primer on the second gene (neo):

```
neo:(3'end of coding region)
5'-...CGC CTT CTT GAC GAG TTC TTC TGA GCG GGA CTC TGG GGT...-3'
3'-...GCG GAA GAA CTG CTC AAG AAG ACT CGC CCT GAG ACC CCA...5'
        |<-------overlap------->|
     3'-GAA GAA CTG CTC AAG AAG A
                               A
                                 C CAC GAC AAC GGA CTA CTT CAG...-5'
```

PCR using this strand and oligonucleotide 'a' as primers produces product AD.

(B) *Generating fragment CF by one-sided SOEing:*

```
ova:
amino acid
   #: 277 278 279 280 281 282 283 284 285 286 287 288 289 290
5'-...AAG ATC AAA GTG TAC TTA CCT CGC ATG AAG ATG GAG GAA AAA...-3'
3'-...TTC TAG TTT CAC ATG AAT GGA GCG TAC TTC TAC CTC CTT TTT...-5'

Primer 'd':           |<--priming region--->| |<--overlap region->|
                   3'-cac atg aat gga gcg tac act cgc cct gag acc c-5'

Product CD (right end):
5'-...AAG ATC AAA GTG TAC TTA CCT GCG ATG TGA GCG GGA CTC TGG G-3'
3'-...TTC TAG TTT cac atg aat gga gcg tac act cgc cct gag acc c-5'
```

The top strand acts as a primer on the neo gene, together with oligonucleotide 'f', to produce CF:

```
5'-...AAG ATC AAA GTG TAC TTA CCT GCG A
                                      T
                                      G
                                        TGA GCG GGA CTC TGG G-3'
                                        |<------overlap------>|
     5'-...CGC CTT CTT GAC GAG TTC TTC TGA GCG GGA CTC TGG GGT...-3'
     3'-...GCG GAA GAA CTG CTC AAG AAG ACT CGC CCT GAG ACC CCA...-5'
```

Figure 8. Details of the primers used for constructing the fusion protein shown in *Figure 7*. This figure shows the sequences involved in the one-sided SOEing reactions which generate fragments AD and CF. The SOEing primers 'c' and 'd' are first used to amplify fragment CD from the *ova* gene. This fragment contains sequences at each end which allow it to overlap with the *neo* gene. Three further reactions are required to generate the product. In SOE reaction #1, shown in part A, the 'overlap' region of primer 'c' allows fragment CD to act as a primer on the *neo* gene along with oligomer 'a' to generate fragment AD. In this reaction, the bottom strand of CD is productive, because it can serve to amplify a segment of *neo* in conjunction with oligomer 'a'. In reaction #2 (part B), fragment CD is again used as a primer on *neo*, but this time with oligo 'd'. In this reaction, it is the top strand of CD which is incorporated into product CF, while the bottom strand is non-productive. The final step is the joining of these two fragments in a two-sided overlap extension reaction. This overlap region is not shown because it is so long.

4.2.2 Choice of restriction enzymes

If the product of a PCR or SOE reaction is to be cloned the most reliable method, in our experience, is to add restriction sites to the flanking primers so that they can be ligated with 'sticky ends'. For this purpose we have obtained better results with some restriction enzymes than with others. This may be due in part to the fact that the ends of a PCR product actually have one strand which is an incorporated synthetic oligonucleotide (see *Figure 4*), and synthetic DNA may be resistant to enzymatic digestion. Another factor which seems to affect how well a restriction enzyme cuts is how close it is to the end of the DNA segment. The following list shows enzymes which have given good results in our hands together with the minimum distance from the end that has worked for us. Shorter distances were not tried, except in the case of *Xho*I as noted.

> *Bam*HI – 3 bp
> *Eco*RI – 4 bp
> *Sal*I – 9 bp
> *Hind*III – 9 bp
> *Xho*I – 20 bp (did not work 10 bp from the end)

These distances should be taken as at best a rough guide. When we say one of the oligos 'did not work' we mean that we were unable to clone a fragment produced with that oligo, even after many attempts, but when the oligo was resynthesized with a much longer 'tail' we had no difficulty. Our supposition is that the enzyme could not cut that close to the end, though this is not proved. In many cases we have adhered to the policy of adding more than one restriction site to the end of the primers, as a back-up. (This adds a longer 'tail' to the inside site, and provides a second site farther out just in case.)

4.3 Establishing a cassette system

We have normally conducted PCR-based manipulations using the 'cassette' approach, in which a small segment or cassette is removed from the cloned target gene, manipulated, and then re-inserted into the slot (see also Chapter 10). This has two advantages; first, smaller fragments are more efficiently amplified by PCR, and second, they make it necessary to sequence a smaller portion of the cloned product.

Cassette sizes in the range of 500 bp are typically the most convenient. This size is easily amplified and sequenced, but is not so small as to be difficult to manipulate. We have, however, cloned SOEn products as large as 1 kb (10) and PCR products smaller than 200 bp with little difficulty (R. M. Horton, manuscript in preparation).

5. Procedures

5.1 Purifying the oligonucleotides

Our oligomers are made on an Applied Biosystems model 380A DNA synthesizer. The final step in the synthesis involves heating at 55°C for 8–15 h in ammonium hydroxide. Dry the oligonucleotides in vacuum (on a SpeedVac, Savant Inc.). Resuspend the dry pellets in 500 μl H_2O and desalt over a Sephadex G-50 column in H_2O. Alternatively, commercially prepared NAP-10 Sephadex G-25 columns (Pharmacia) may be used. Collect 500 μl fractions and assay for absorbance at 260 nm on a UV spectrophotometer. Pool together the fractions containing the first peak (typically fractions 3, 4, and 5 from a NAP-10 column) and measure their absorbance at 260 nm to determine the concentration (see Section 5.2).

Immediately after purification the oligonucleotides should be divided into small aliquots some of which are held in reserve in case the working stocks become contaminated with a template gene.

5.2 Determining the amount of primer to add

For our standard 100 μl reaction, we want to add 100 pmol of each primer to give a concentration of 1 pmol/μl, or 1 μM. The units used here can be confusing for those who do not routinely work with such small quantities, so we will go through this in detail. The following SI prefixes will be used:

$$\text{micro- } (\mu) = 10^{-6}.$$
$$\text{nano- } (n) = 10^{-9}.$$
$$\text{pico- } (p) = 10^{-12}.$$

For oligonucleotides the concentration is 20 μg/ml (equivalent to 20 ng/μl) per unit of absorbance at 260 nm (see ref. 11). Thus, multiplying the absorbance units by 20 will give the oligonucleotide concentration in ng/μl. The average molecular weight of a nucleotide is approximately 310 g/mol. This is equivalent to 31 ng/100 pmol. For an oligomer, then, 100 pmol will weigh 31 ng per base. If we call the concentration 'Y' ng/μl, the length of the oligomer 'N' bases, and the volume of primer we want to add per reaction 'Z' μl/100 pmol, then this is calculated by

('N' bases) × (31 ng/100 pmol.base) × (1 μl/'Y'ng) = 'Z' μl/100 pmol.

It may be convenient to dilute some of the oligonucleotide so that its concentration is 10 pmol/μl (that is, the same as 10 μM), so that 10 μl would be added to a standard 100 μl reaction.

5.3 Parameters affecting PCR and SOE reactions

The first step (and occasionally the most frustrating) is to get the initial PCR(s) to work. The rule about which parameters affect PCR is simple: **everything**. However, it is our experience that the most critical parameter to vary in order to optimize the reaction is the concentration of Mg^{2+}.

5.3.1 Magnesium concentration

The standard reaction buffer (*Table 1*) uses 1.5 mM $MgCl_2$, but in cases where this does not work well we titrate this concentration over a range from 0.5 mM to 2.5 mM in 0.5 mM increments. We find that one of these concentrations usually works well. In general, higher Mg^{2+} concentrations tend to produce higher backgrounds of undesired products, and there is usually a trend toward less and less background as the Mg^{2+} concentration is reduced until at some concentration only the band of interest is left. However, the band of interest may appear at a low Mg^{2+} concentration with no bands at all appearing at a slightly higher concentration. Also, some background bands may persist at all Mg^{2+} concentrations. In these situations other workers advise titrating primer concentration, dNTP concentration, and other parameters. However, we have usually been able to find a Mg^{2+} concentration which worked well enough to give us sufficient product to proceed without fine-tuning all of the other variables. Since higher free Mg^{2+} concentrations lead to more errors by *Taq* polymerase, the lowest workable Mg^{2+} concentration is preferred. A study of effects of various parameters on *Taq* error rate has recently appeared (12).

5.3.2 Template concentration

Theoretically, starting with large amounts of template means that fewer rounds of replication are required to generate a given amount of PCR product, and this lessens the opportunity for the polymerase to misincorporate bases. With cloned templates a very wide variety of template concentrations usually works up to about the microgram range. With very high template concentrations, however, the reaction is sometimes impeded. Also, when very few rounds of PCR are performed (e.g. by starting with a large number of template molecules), the fraction of 'open-ended' strands (strands which extend beyond the ends of the primers) is higher. It is possible that these largely single-stranded molecules contribute to the background of unwanted products which is observed when the intermediates are not purified by size-selection (4).

5.3.3 Construct complexity

The complexity of the construction can also influence the error frequency, probably by changing the total number of rounds of replication required. For example, in a two-sided mutagensis project involving joining two fragments by overlap extension, an error frequency of 0.026% was observed, as mentioned previously. In more complicated constructs which involved joining four separate fragments two at a time, so that each segment of the final product had undergone one PCR and two sets of SOE reactions, the error frequency was approximately 0.06%.

5.3.4 Reaction volume

Finally, we have noticed cases in which a reaction volume smaller than $100\,\mu l$ (50 or $25\,\mu l$/tube) actually produced more product. We suspect that sometimes heat transfer is a problem with the larger volumes.

5.4 Purifying the intermediate products

PCR products which are to be joined by overlap extension should be purified, as this cuts down on the generation of unwanted byproducts (4). Other workers use less exacting purification schemes (3), but our experience is that size purifying the intermediates through a gel leads to more efficient formation of the final product. A simplification of the technique of one-sided SOEing has been reported for recombining unrelated sequences in which the intermediates are not purified at all and the entire procedure is carried out in one tube (13). The reaction includes two template genes and primers 'a', 'b', and 'c'. Primers 'a' and 'b' cause product AB to be amplified from the first template. Primer 'b' contains sequences which allow product AB to overlap on the second template to form the recombinant product AD. By including primer 'b' at a low concentration (about $0.01\,\mu M$), only a small amount of product AB is produced. Otherwise, it might compete for reactants and decrease the yield of recombinant. The exact amount of primer 'b' which gives optimal yield may vary between systems, and it may be necessary to perform a titration. In our limited experience with this one-tube technique we have found that these reactions produce more 'artefact' bands and less of the desired product than when the intermediates are gel-purified. Also note that this modification would not be practical for site-directed mutagenesis unless some way were found to prevent amplification of the parental sequence by primers 'a' and 'b'.

Normally, we cut out the band of the correct size and recover the DNA from the gel by adhesion to GlassMilk using a GeneClean kit (Bio 101). For the GeneClean procedure electrophoresis buffers containing borate cannot be used, so we use Tris-acetate buffer (TAE, ref. 9, p. 154). However, the GeneClean procedure is only useful for fragments bigger than about 200 bp. For smaller fragments we have used electroelution either by cutting a well in the gel in front of the band and running the fragment into it (see ref. 11, p. 167), or by cutting the band out and electroeluting in dialysis tubing (ref. 11, p. 164). The electroeluted DNA is then precipitated with alcohol and potassium acetate, resuspended in water, and used in the subsequent reaction.

Recently, we have begun using a modified 'freeze-squeeze' method to recover DNA from gel slices (14). This procedure, illustrated in *Figure 9*, involves freezing the agarose band and centrifuging out the liquid, including the DNA. The DNA is recovered in a small enough volume that it can be used directly, and does not need to be precipitated. This method is faster, easier, and gives greater recovery than other methods we have used. However, if no alcohol precipitation or washes are performed, there will probably be a certain amount of ethidium bromide carried over into the subsequent

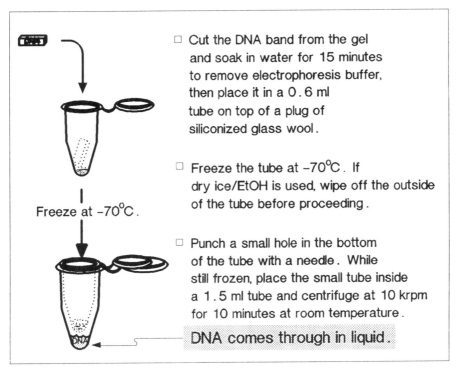

Cut the DNA band from the gel and soak in water for 15 minutes to remove electrophoresis buffer, then place it in a 0.6 ml tube on top of a plug of siliconized glass wool.

Freeze the tube at –70°C. If dry ice/EtOH is used, wipe off the outside of the tube before proceeding.

Punch a small hole in the bottom of the tube with a needle. While still frozen, place the small tube inside a 1.5 ml tube and centrifuge at 10 krpm for 10 minutes at room temperature. DNA comes through in liquid.

Freeze at –70°C.

Figure 9. The 'freeze-squeeze' method for recovering DNA from a slice of an agarose gel (see ref. 11).

reaction. We do not at present know how significantly, if at all, this affects the random error rate.

Small fragments are resolved in higher percentage agarose gels. For percentages greater than 1% we use a mixture of 'regular' agarose (SeaKem LE, FMC BioProducts) and high concentration agarose (NuSieve GTG, FMC BioProducts). In these mixtures 1% of the agarose is the regular type and the remainder (up to a total of at most 4%) is NuSieve.

5.5 Cloning PCR and SOEn products

For many applications the product generated with PCR technology must be cloned. This cloning step is like any cloning step, and any protocol that the reader is comfortable with is probably adequate, but we include ours in *Protocol 3* for completeness. It is different from some protocols in that we attempt to limit the use of phenol extractions which we view as having great potential for losing product and which may decrease the efficiency of ligations. For products greater than about 250 bp we use GeneClean (Bio 101) to purify the fragments before ligation because it is easy and gets the DNA very clean, which increases ligation efficiency. Smaller fragments are not easily recovered from GeneClean, so we usually resort to some method involving phenol extraction and alcohol precipitation in those cases.

The checking steps are very important because if anything goes wrong they help you to immediately identify the problem. One checking gel can be poured on the first day, and then it can be wrapped in plastic wrap and saved to be used the next day to check the ligation.

By adding the vector to the PCR product before manipulating it, we can monitor the vector for cutting instead of having to examine the PCR product. (If the cloning sites are in the PCR primers, digestion with restriction enzymes will not change the size enough to tell on an agarose gel). This is not an option for systems in which a cassette is to be replaced by the same-sized modified fragment. In those cases the vector must be cut separately and purified away from the parental type insert before it is used.

Protocol 3. Cloning PCR generated fragments

A. *If the vector does not already contain an insert:*

1. Amplify enough material to work with. (Approximately 1 to 4 μg. A 100 μl reaction should produce enough if it works reasonably well.)

2. Run a checking gel and *roughly* estimate the amount of product band, by comparing its intensity to the marker. Add 10 μg of clean yeast tRNA (BRL) as a carrier. Add enough vector (plasmid or M13) to give approximately a 2:1 insert:vector molar ratio. (For 5 μg of a 1.5 kb PCR fragment, you would add 0.5 μg of M13, which is about 7.2 kb.) Extract once with chloroform/isoamyl alcohol to remove the oil.

3. Ethanol precipitate the insert plus vector mixture:

   (a) Add 2.5 vol of 2% KOAc in 95% EtOH.

   (b) Spin in microcentrifuge 30 min at 12 000 r.p.m. at 4°C.

   (c) Carefully remove supernatant without disturbing the pellet (the pellet should be visible, though it will be small).

   (d) Rinse the pellet with cold 70% EtOH. Remove as much EtOH as possible, and then spin briefly in a microcentrifuge to remove residual EtOH from the walls of the tube. Remove the last drops of EtOH with a small pipette tip.

   (e) Dry the pellet in a speed vac or lyophylizer.

   Note: If the PCR was not optimal, and many other bands are present besides the band of interest, it may be advisable to purify the appropriate band on a gel before proceeding. The insert can be digested with the appropriate restriction enzyme(s) before this purification, as then it only has to be cleaned thoroughly one time before ligation.

4. Resuspend in 90 μl of water. Add 10 μl of the proper 10 × restriction enzyme buffer and an excess of enzyme. This large reaction volume helps to dilute any remaining salts in the pellet. For most enzymes add about 8 U. For enzymes with star activity (i.e. *Bam*HI and *Eco*RI), use about 3 U. Incubate at the appropriate temperature for several hours to overnight.

5. Check digestion by running 10 μl on a checking gel. You should be able to see that the vector is cut. If it is not, add 5 U more enzyme and incubate again. *Be sure the vector is completely cut before going on to the next step.*

6. Recover the DNA from the cutting buffer by using GeneClean (BIO 101). Add 2.5 vol of NaI solution (approximately 250 μl) directly to the solution containing the DNA. Use 5 μl GlassMilk and follow the directions in the kit. Elute the DNA in a total volume of 10 μl (2 × 5 μl) of H$_2$O.

Note: for small inserts, GeneClean is not recommended, because the small fragments bind irreversibly to the GlassMilk. Thus, for small inserts (less than about 150–200 bp), phenol extraction and ethanol precipitation are recommended.

7. Add 1.1 μl of 10× ligase buffer. Remove 1 μl of this mixture and save for the checking gel. Add 0.5 μl of ligase and incubate overnight at 4°C.

8. Run a checking gel of the unligated samples saved from step 7, and 1 μl of the ligation reaction. You should see a difference between them if the ligation worked, and if you recovered anything with GeneClean (you should see more bands in the ligated lane, representing concatamers, closed circles, recombinant products, etc.). If you see no evidence of ligation, add more ligase and incubate again.

9. Transform the appropriate host bacteria and select transformants.

B. *If the vector already contains an insert similar to the modified sequence the above protocol should be modified as follows*:

Step 2: Do not add the uncut vector to the PCR product.

Step 3: EtOH precipitate the PCR mixture as above.

Step 4: Cut with restriction enzyme as above. In a separate tube, cut the vector.

Step 5: Check digestion of the vector. If your restriction sites are in the primers for the PCR product, you will not be able to see if it cut anyway, so there is no use checking.

Step 6: Run preparative amounts of the cut vector and cut insert on an agarose gel. Cut out the vector band, and the band containing the modified insert, being careful not to contaminate either one with the parental-type insert band. Recover the insert and vector bands from the gel. Here it is

Protocol 3. *continued*

important that the DNA be very clean, because it is about to be ligated. We use GeneClean to both recover the DNA and to get it clean enough to ligate. Visually estimate the quantity of insert and vector recovered on a checking gel, and mix them to achieve approximately a 2:1 insert:vector ratio.

Steps 7 & 8: Ligate and transform as above.

5.6 Sequencing the cloned products

Due to the possibility of random errors being made by *Taq* polymerase it is advisable, for most applications, to sequence the product. In some cases (for example, if there is some functional assay for the product) it may not be important to have an error-free clone. Or if the entire population of PCR product is used in an experiment (perhaps for a gel shift assay), the over-whelming majority of the molecules will be correct, and there is no need to sequence such a population. In our applications it has been necessary to have error-free clones, and we have generally used double-stranded sequencing directly from our expression vector using a Sequenase kit (USB) with a modified protocol (15).

6. Discussion

6.1 Other applications

We have largely focused on specific manipulation of small segments of genes which are then cloned into a plasmid for study. Third generation technologies actually have a tremendously broad range of potential applications, some of which we discuss below.

6.1.1 Direct use of PCR generated fragments

Many applications require small fragments of DNA not contained in a vector and for the production of these fragments PCR manipulation is ideal. The small rate of error found in products amplified from cloned templates means that the percentage of molecules containing any given random mutation is very small and can be ignored for many applications which employ the entire population of PCR generated fragments. Such applications would include *in vitro* transcription (3) and translation (5), and perhaps gel retardation assays. This should prove a very convenient method for testing the effects of site-directed mutations on functions which can be assayed on the entire PCR generated population.

6.1.2 Random mutagenesis

By varying the reaction conditions the frequency of random errors generated can be influenced by a variety of factors (e.g. more rounds of replication lead to more mutations). This may prove useful for generating random mutations in a stretch of DNA (see section 5.1.3 and ref. 12).

It has been reported that *Taq* polymerase may add random sequences to the 3'-ends of strands it generates. These random mutations do not normally show up as insertions near the joints in SOEn products, however. This is expected, because a mismatch at the 3'-end of a DNA strand will prevent it from acting as a primer (16).

6.1.3 Random recombination

Recombination between related genes being amplified in the same reaction has been reported (17). This is probably due to partially extended strands of one gene being able to overlap and extend on the other gene in subsequent rounds. While this may be a problem when one is trying to amplify a gene out of a multigene family, it also may provide a convenient way to generate sets of random recombinants between homologous genes. The frequency of random recombination may be higher when the template DNA is in bad shape. For example, in the amplification of ancient DNA from museum or archaeological specimens, which has accumulated extensive damage over time in the form of oxidation, lost bases, cross-links, and so on, the process of 'jumping PCR' is thought to be especially important (18). This implies that, if one wanted to cause random recombination in a PCR, it might be useful to damage the template DNA first, perhaps by treatment with an oxidizing agent such as hydrogen peroxide.

6.2 Concepts for further development

The third generation of recombinant DNA technology is just beginning, and there is a lot of potential yet to develop. Here we outline some ideas which may be useful, or at least interesting, though they have not yet been developed as working protocols.

6.2.1 SOEing directly into a vector

The procedures outlined in this chapter are actually partly second generation technology, since we use restriction enzymes and ligase to insert the product into a vector for cloning. This makes development of a cassette system necessary, and it would be convenient if this could be avoided. It may be possible to SOE a fragment directly into a vector and have an entirely third generation system. We describe three approaches to this problem.

i. Blunt-end recircularization
The vector and insert are SOEn together into a linear molecule and then circularized by blunt-end ligation. The feasibility of all the steps necessary for this approach has been shown. An entire plasmid can be amplified by PCR (e.g. ref. 19) and a PCR generated fragment has been SOEn to a linearized plasmid (Sandhu, Precup, and Kline, pers. commun.). These plasmid con-

structs can then be recircularized by blunt-end ligation and used to transform bacteria.

ii. Overlapping fragments to produce a nicked circle

This idea is illustrated in *Figure 10A*. It relies on the fact that the host bacterium will repair single-stranded nicks in a plasmid, as long as the overlaps between the nicked strands are long enough to be stable. This approach would require no ligation *in vitro* yet would potentially allow a large fragment to be constructed from smaller, PCR generated parts. An idea similar to this has recently been reported (20).

iii. Recircularization into a 'phragmid'

The lambda-Zap cloning system (Stratagene) uses a 'phagemid' system to excise a circular replicon from a linear lambda phage derivative (21). This system takes advantage of the fact that an M13-like replication origin can be split into initiation and termination parts to permit rolling circle replication from a linear template. *Figure 10B* illustrates how this approach might be applied to circularize a linear fragment generated by gene SOEing. Since this system involves a PCR fragment including plasmid and phage replication origins, it might be called a 'phragmid'.

6.2.2 Recombining larger fragments

The size of the fragments which can be manipulated by PCR-based methods is affected by the size of the fragments which can be amplified. Below are some ideas which might be useful for working with longer fragments. Any developments for lowering the error frequency of PCR (such as a less error-prone thermostable polymerase) will also make working with longer segments more practical.

i. The long oligo/short oligo scheme

This is illustrated in *Figure 10C*. In this idea, two oligos are synthesized for each of the ends to be recombined, one of which is longer than the other. The products are then mixed, denatured, and reannealed, and fragments with sticky ends are generated, a certain percentage of which are compatible. It may be possible with this scheme to overcome limitations of the size of the product which can be generated efficiently. Compatible sticky ends as short as two base-pairs are routinely ligated when restriction sites are used. This method would allow the generation of much longer sticky ends, which would be more stable. A similar idea has recently been demonstrated (22).

ii. Directed ligation

An alternative third generation technology could be based on the ligation amplification reaction (23, 24) which uses a fragment of DNA to direct the

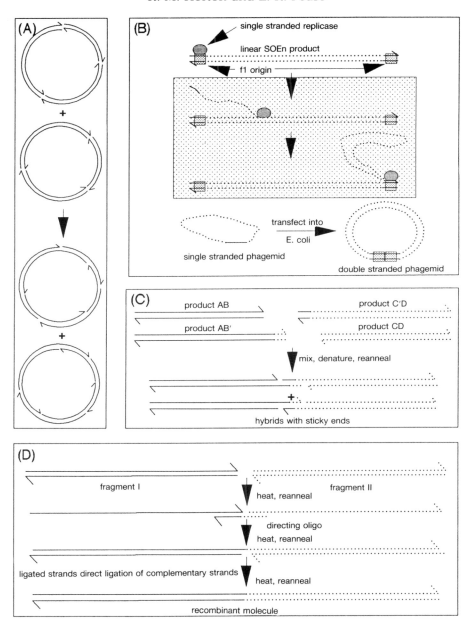

Figure 10. Conceptual applications of third-generation technology. These ideas are discussed in Section 6.2.

ligation of two synthetic oligonucleotides. If this idea were applied in the other direction (that is, to use a synthetic oligo to direct ligation of two DNA fragments as shown in *Figure 10(D)*), it would possibly allow precise recombination of sequences. This approach would not require the resynthesis of the entire recombinant product *in vitro*, as gene SOEing does. It thus has the potential advantages of being less limited in terms of the size of the fragments which can be recombined and of not being prone to replication errors. Such an application of directed ligation may prove very efficient, because by repeating the process each ligated product can serve to direct the ligation of the complementary strands. Also, it should be possible to direct the ligation of products into a circle, to make a plasmid. In such an approach PCR would prove useful for generating the fragments to be ligated because it allows the ends to be easily and precisely defined at essentially any point along the genes. [We have recently tried this idea and have run into technical problems. Anyone interested in this approach should feel free to contact the authors].

6.2 PCR: the technology for most applications

PCR has become a core technique for molecular biologists in an astonishingly short time. It has a tremendous variety of applications. PCR can detect DNA sequences with a sensitivity far beyond that of hybridization techniques, such as Southern blotting. It can be used to detect cDNA sequences with greater sensitivity than Northerns. It can be used to perform site-directed mutagenesis and can save tremendous effort in screening recombinant bacteria, plasmids, and phage. It is in effect an all around technique and to a certain degree it simplifies the technical expertise required to perform molecular biology. Since PCR can be used to precisely recombine DNA sequences without reliance on restriction sites, many complicated recombinant DNA problems can be solved in a straightforward manner without having to introduce, eliminate, or modify restriction sites, and without having to perform unreliable procedures, as blunt-end ligation appears to be for many people. Expertise in PCR technology is therefore helpful in a tremendous number of experimental problems involving both synthesis and analysis of genes.

The birth of a new generation technology is an exciting event, and it can be quite rewarding to help it along. There are tremendous opportunities for utilizing PCR-based approaches, and much room for developing new ones. We hope that we have provided some appreciation of the potential of this technology for manipulating genetic sequences, and encouragement to our colleagues to develop them further.

Acknowledgements

R.M.H. is supported by a pre-doctoral fellowship from the Mayo Graduate School, Mayo Foundation. We are very grateful to Drs Henry Hunt, Jeff

Pullen, Zeling Cai, Gurpreet Sandhu, Jim Precup, and (especially) Gobinda Sarkar for sharing their results and/or protocols before publication, and to Steffan Ho for helpful discussions.

References

1. Mullis, K. and Faloona, F. (1987). In *Methods in Enzymology* (ed. R. Wu), Vol. 155, pp. 335–50.
2. Kadowaki, H., Kadowaki, T., Wondisford, F. E., and Taylor S. I. (1989). *Gene*, **76**, 161.
3. Higuchi, R., Krummel, B., and Saiki, R. K. (1988). *Nucleic Acids Research*, **15**, 7351.
4. Ho, S. N., Hunt, H. D., Horton, R. M., Pullen, J. K., and Pease, L. R. (1989). *Gene*, **77**, 51.
5. Sarkar, G. and Sommers, S. S. (1990). *BioTechniques*, **8**, 404.
6. Perrin, S. (1990). *Nucleic Acids Research*, in press.
7. Lawyer, F. C., Stoffel, S., Saiki, R. K., Myambo, K., Drummond, R., and Gelfand D. H. (1989). *Journal of Biological Chemistry*, **264**, 6427.
8. Wu, R. and Grossman, L. (eds.). (1987). *Methods in Enzymology*, Vol. 154, pp. 329–430.
9. Suggs, S. V., Hirose, T., Miyake, T., Kawashima, E. H., Johnson, M. J., Itakura, K., and Wallace, R. B. (1981). *Developmental Biology Using Purified Genes* (ed. D. D. Brown and C. F. Fox), pp. 683–93). Academic Press, New York.
10. Horton, R. M., Hunt, H. D., Ho, S. N., Pullen, J. K., and Pease, L. R. (1989). *Gene*, **77**, 61.
11. Maniatis, T., Fritch, E. F., and Sambrook, J. (ed.). (1982). *Molecular Cloning: A Laboratory Manual*, p. 468. Cold Spring Harbor Laboratory, Cold Spring Harbor, New York.
12. Eckert, K. A. and Kunkel, T. A. (1990). *Nucleic Acids Research*, **18**, 3739.
13. Yon, J. and Fried, M. (1989). *Nucleic Acids Research*, **17**, 4895.
14. Tautz, D. and Renz, M. (1983). *Analytical Biochemistry*, **132**, 14.
15. Kraft, R., Tardiff, J., Krauter, K. S., and Leinwand, L. A. (1988). *BioTechniques*, **6**, 544.
16. Newton, C. R., Graham, A., Heptinstall, L. E., Powell, S. J., Summers, C., Kalsheker, N., Smith, J. C., and Markham, A. F. (1989). *Nucleic Acids Research*, **17**, 2503.
17. Shuldiner, A. R., Nirula, A., and Roth, J. (1989). *Nucleic Acids Research*, **17**, 4409.
18. Paabo, S., Higuchi, R. G., and Wilson, A. C. (1989). *Journal of Biological Chemistry*, **264**, 9709.
19. Mole, S. E., Iggo, R. D., and Lane, D. P. (1989). *Nucleic Acids Research*, **17**, 3319.
20. Shuldiner, A. R., Scott, L. A. and Roth, J. (1990). *Nucleic Acids Research*, **18**, 1920.
21. Short, J. M., Fernandez, J. M., Sorge, J. A., and Huse, W. D. (1988). *Nucleic Acids Research*, **16**, 7583.
22. Jones, D. H. and Howard, B. H. (1990). *BioTechniques*, **8**, 178.
23. Wu, D. Y. and Wallace, R. B. (1989). *Genomics*, **4**, 560.
24. Landegren, U., Kaiser, R., Sanders, J., and Hood, L. *Science*, **241**, 1077.

A

Suppliers of specialist items

Aldrich Chemical Company, Milwaukee, WI, USA; The Old Brickyard, New Road, Gillingham, Dorset SP8 4JL, UK.

Amersham International, Amersham Place, Little Chalfont, Buckinghamshire HP7 9NA, UK; 2636 S. Clearbrook Drive, Arlington Heights, IL 60005, USA.

Amicon Corporation, 17 Cherry Hill Drive, Danvers, MA 01923, USA; Upper Mill, Stonehouse, Gloucester GL10 2J, UK.

Applied Biosystems Inc., 850 Lincoln Center Drive, Foster City, CA 94404, USA; Kelvin Close, Birchwood Science Park North, Warrington, Cheshire WA3 7PB, UK.

Beckman Instruments, 2500 Harbor Boulevard, PO Box 3100 Fullerton, CA 92634, USA.

Bio101 Inc., Box 2284, La Jolla, CA 92038-2284, USA; Stratech Scientific Ltd., 61/63 Dudley Street, Luton, Bedfordshire LU2 0NP, UK.

Bio-Rad, 1414 Harbour Way South, Richmond, CA 94804, USA; Claxton Way, Watford Business Park, Watford, Hertfordshire WD1 8RP, UK.

Boehringer Mannheim GmbH, Postfach 310120, D-6800 Mannheim 31, FRG; PO Box 50816, Indianapolis, IN 46250, USA.

Brinkmann Instruments, Cantiague Road, Westbury, NY 11590, USA.

BRL, *see* Gibco BRL.

Calbiochem, 10933 North Torrey Pines Road, La Jolla, CA 92037, USA.

Cetus, *see* Perkin Elmer Cetus.

Cruachem Ltd., Todd Campus, West of Scotland Science Park, Acre Road, Glasgow G20 0UA, UK; Herndon, VA, USA.

Diagen GmbH, Neiderheider Strasse, 4000 Dusseldorf 13, FRG; Qaigen Inc., Studio City, CA 91640, USA.

DuPont Company, Biotechnology Systems Division, BRML, G-50986, Wilmington, DE 19898, USA; Wedgewood Way, Stevenage, Hertfordshire SG1 4QN, UK.

Fisher Scientific, Pittsburgh, PA, USA; Zurich, Switzerland.

Fluka Chemical Co., 980 S. Second Street, Ronkonkoma, NY 11779, USA; Peakdale Road, Glossop, Derbyshire SK13 9XE, UK.

Gibco-BRL, Grand Island, NY, USA; PO Box 35, Trident House, Renfrew Road, Paisley PA3 4EF, UK.

Mallincrodt Inc., Partis, KY, USA.

Marine Collioids, FMC Corporation, Bioproducts Division, 5 Maple Street, Rockland, ME 04841, USA; 1 Risingevej, DK-2665 Vallensbaek Strand, Denmark.

Millipore Corp., 80 Ashby Road, Bedford, MA 01730, USA; The Boulevard, Blackmoor Lane, Watford, Hertfordshire WD1 8YW, UK.

New England Biolabs, 32 Tozer Road, Beverley, MA 01915-9990, USA; Postfach 2750, 6231 Schwalbach/Taunus, FRG.

New England Nuclear, *see* DuPont.

Perkin Elmer Cetus, Main Avenue; Norwalk, CT 06859-0012, USA; Postfach 101164, 7770 Ueberlingen, FRG; Maxwell Road, Beaconsfield, Buckinghamshire HP9 1QA, UK.

Pharmacia LKB Biotechnology AB, S-75182 Uppsala, Sweden; 800 Centennial Avenue, Piscataway, NJ 08854, USA.

Promega Corp., 2800 S. Fish Hatchering Road, Madison, WI 53711, USA.

Savant Instruments Inc., Farmingdale, NY, USA.

Sigma Chemical Co., PO Box 14508, St. Louis, MS 63178, USA; Fancy Road, Poole, Dorset BH17 7NH, UK.

Stratagene, 11099 North Torrey Pines Rd., La Jolla, CA 92037, USA; Postfach 105466, D-6990 Heidelberg, FRG; Cambridge Innovation Centre, Cambridge Science Park, Milton Road, Cambridge CB4 4GF, UK.

United States Biochemical Corp., PO Box 22400, Cleveland, Ohio 44122, USA; Cambridge Bioscience, Newton House, 42 Devonshire Road, Cambridge CB1 2BL, UK.

Whatman BioSystems Ltd, Springfield Mill, Maidstone, Kent ME14 2LE, UK; 9 Bridewell Place, Clifton, NJ 07014, USA.

Worthington, Freehold, NJ 07228, USA; Flow Laboratories Ltd, P.O. Box 17, Second Avenue, Industrial Estate, Irvine KA12 8NB, UK.

Index